SpringerBriefs in Petroleum Geoscience & Engineering

Series Editors

Dorrik Stow, Institute of Petroleum Engineering, Heriot-Watt University,
Edinburgh, UK

Mark Bentley, AGR TRACS International Ltd, Aberdeen, UK

Jebraeel Gholinezhad, School of Engineering, University of Portsmouth,
Portsmouth, UK

Lateef Akanji, Petroleum Engineering, University of Aberdeen, Aberdeen, UK

Khalik Mohamad Sabil, School of Energy, Geoscience, Infrastructure and Society,
Heriot-Watt University, Edinburgh, UK

Susan Agar, Oil & Energy, Aramco Research Center, Houston, USA

Kenichi Soga, Department of Civil and Environmental Engineering, University of
California, Berkeley, USA

A. A. Sulaimon, Department of Petroleum Engineering, Universiti Teknologi
PETRONAS, Seri Iskandar, Malaysia

The SpringerBriefs series in Petroleum Geoscience & Engineering promotes and expedites the dissemination of substantive new research results, state-of-the-art subject reviews and tutorial overviews in the field of petroleum exploration, petroleum engineering and production technology. The subject focus is on upstream exploration and production, subsurface geoscience and engineering. These concise summaries (50–125 pages) will include cutting-edge research, analytical methods, advanced modelling techniques and practical applications. Coverage will extend to all theoretical and applied aspects of the field, including traditional drilling, shale-gas fracking, deepwater sedimentology, seismic exploration, pore-flow modelling and petroleum economics. Topics include but are not limited to:

- Petroleum Geology & Geophysics
- Exploration: Conventional and Unconventional
- Seismic Interpretation
- Formation Evaluation (well logging)
- Drilling and Completion
- Hydraulic Fracturing
- Geomechanics
- Reservoir Simulation and Modelling
- Flow in Porous Media: from nano- to field-scale
- Reservoir Engineering
- Production Engineering
- Well Engineering; Design, Decommissioning and Abandonment
- Petroleum Systems; Instrumentation and Control
- Flow Assurance, Mineral Scale & Hydrates
- Reservoir and Well Intervention
- Reservoir Stimulation
- Oilfield Chemistry
- Risk and Uncertainty
- Petroleum Economics and Energy Policy

Contributions to the series can be made by submitting a proposal to the responsible Springer contact, Charlotte Cross at charlotte.cross@springer.com or the Academic Series Editor, Prof Dorrik Stow at dorrik.stow@pet.hw.ac.uk.

More information about this series at http://www.springer.com/series/15391

Vahid Tavakoli

Carbonate Reservoir Heterogeneity

Overcoming the Challenges

 Springer

Vahid Tavakoli
School of Geology, College of Science
University of Tehran
Tehran, Iran

ISSN 2509-3126   ISSN 2509-3134   (electronic)
SpringerBriefs in Petroleum Geoscience & Engineering
ISBN 978-3-030-34772-7   ISBN 978-3-030-34773-4   (eBook)
https://doi.org/10.1007/978-3-030-34773-4

This Springer imprint is published by the registered company Springer Nature Switzerland AG
The registered company address is: Gewerbestrasse 11, 6330 Cham, Switzerland

Preface

Heterogeneity is an intrinsic property of all carbonate reservoirs. The properties of these reservoirs considerably change both laterally and vertically. Lateral changes are usually the result of various depositional settings, while vertical heterogeneities are caused by basin evolution through time. Available data from these reservoirs are very limited and so predicting the distribution of properties between the wells is very complicated. The hydrocarbon in place is calculated using these predictions and so they are very important. In fact, many aspects of the reservoir studies are about heterogeneities. Facies analysis and classifications, determining sedimentary environments, reservoir rock typing, flow unit determination and sequence stratigraphy are some examples.

Despite this importance, few studies have been published about the heterogeneity of the carbonate reservoirs. How the heterogeneity of a reservoir is evaluated? Where we should start and how the process continues? How the various scales of heterogeneity are related to each other? This book tries to answer these questions. It starts with an introduction about the heterogeneity and states the problem. The term is defined, and the importance of its study is explained. Then, its causes are considered, and the required materials are discussed. Chapter 1 ends with the scales of heterogeneity which are one of the most important aspects of this concept. This book has been organized based on the scale of heterogeneity from micro- to macroscale. Each chapter is divided into two parts. At first, the problem is discussed, and then, the solution is considered. In Chap. 2, facies analysis, diagenetic impacts on the reservoir, porosity–permeability relationships and pore throat sizes are considered, and then, pore system classifications, rock typing and electrofacies are introduced to overcome the challenge of heterogeneity in microscale. Chapter 3 is about the heterogeneities in mesoscale. Sedimentary environments are defined to organize the homogenous facies in a larger body. Hydraulic flow units, cyclicities and stratigraphic correlations are discussed for the same reason. In macroscopic scale which is the subject of Chap. 4, larger-scale variables such as seismic data and interpretations, fracturing, stratifications, sequence stratigraphic concepts and maps are discussed. This book ends with the petrophysical evaluations and how the analyst can overcome the challenges of heterogeneities in these calculations.

I am thankful to my beloved wife, Dr. M. Naderi-Khujin, for designing most figures of this book. Her abilities in illustrating the most complicated concepts with some simple but beautiful figures are endless. Our personal communications also helped me a lot, especially on macroscopic heterogeneities. I want to thank my colleague Dr. H. Rahimpour-Bonab for giving me brilliant ideas about different aspects of reservoir heterogeneity. I also appreciate the cooperation of my Ph.D. student A. Jamalian for illustrating some figures and helping me in organizing some parts of this book. The preparation of this book was made possible through the help of my M.S. students, M. Nazemi, M. H. Nazari, A. Mondak and B. Meidani. Last but not least, I want to thank all of my colleagues and students at the University of Tehran which are really trying to develop the science of geology in any way that they can. Please let me know any idea, critique or suggestion which you probably have about this book.

Tehran, Iran   Vahid Tavakoli
September 2019

Contents

Chapter 1
Reservoir Heterogeneity: An Introduction

Abstract The ultimate goal of reservoir studies is predicting the properties and their controlling factors in a reservoir body based on limited data mainly from boreholes. Reservoir heterogeneity means variations of reservoir properties in space and time, and so this concept is the most important factor in reservoir studies. Despite such importance, relatively few works have been focused or documented different aspects of this subject. Evaluating heterogeneity is more complicated in carbonates which have diverse facies and are more prone to diagenetic processes. Textures and allochems vary considerably at small scales of carbonate reservoirs. Different facies are deposited in various depositional environments. They also change in response to sea-level changes and climate conditions. These building blocks integrate to create various facies belts and geometries of the depositional settings. These geometries are used for propagation of reservoir properties in field scale. Diagenetic processes modify these properties. They follow the primary textural characteristics in many cases, especially in early diagenesis. Anyway, many late diagenetic processes, such as fractures, crosscut the primary facies as well as other diagenetic features. While facies variations create heterogeneity at larger scales, diagenesis is responsible for changes at smaller scales.

Heterogeneities are present in microscopic, mesoscopic, macroscopic and megascopic scales. Facies and pore types are examples of microscopic variations, while sedimentary structures, stratification and reservoir compartmentalization are considered as large-scale heterogeneities. They can be studied using various tools and methods from thin sections to core CT scanning, core description, reservoir zonation, correlations, maps and seismic sections.

1.1 Definition

In the simplest approach, heterogeneity means diversity in properties in a single body. The terms variability, discrepancy, randomness, complexity, diversity and deviation from a norm could be compared with this concept. In reservoir studies, the properties of interest are fluid flow and storage-related characteristics. All reservoir-related variabilities are included in this concept. Common examples include mineralogy, absolute and relative permeability, pore types, pore volume, pore throat size distribution, textural properties such as grain size and sorting, diagenetic processes such as cementation and dissolution and the production rate. Some authors (Li and Reynolds 1995; Zhengquan et al. 1997) believe that the change in property in three-dimensional space causes system heterogeneity, while others (e.g., Fitch et al. 2015) add the time concept to the definition. They stated that changing a property over time also increases the heterogeneity of a system. Change in discrete object density in space was also defined as heterogeneity (Frazer et al. 2005).

The heterogeneity of a sample or volume strongly depends on the property of interest. These properties could be dependent on or independent from each other. For example, one meter of reservoir core in well scale could be homogeneous from porosity frequency distribution point of view, while it is heterogeneous when pore types are considered. Permeability distribution in this core may be dependent on or independent from porosity. For uniform interparticle porosities, for example, it depends on porosity, but in the case of various pore types, it is mainly independent from pore volume.

Based on previous literatures, Fitch and his co-workers (2015) defined heterogeneity as the change in one or a combination of some various parameters in space and/or time which strongly depends on desired scale. They correctly point to the concept of a combination of parameters which determines the reservoir quality variation in space and time. For many years, researchers tried to define a unique unit which includes all of these properties. The term rock type is commonly used for this purpose. Although various methods have been introduced so far (such as geological rock typing (GRT) (Tavakoli 2018), reservoir quality index (RQI) and flow zone indicator (FZI) (Amaefule et al. 1993), Winland R_{35} method (Kolodzie 1980), Lucia rock fabric number (RFN) (Lucia and Conti 1987; Lucia 1995), none of these methods can be applied in all cases. It means that more research is necessary in this field.

Weber (1986) defined heterogeneity as non-uniform variations of reservoir parameters such as porosity, permeability and pore types in space. He believed that the heterogeneity of a reservoir is initially caused by primary depositional characteristics (facies), followed by diagenetic processes after deposition. This is true because all reservoir properties are the result of these two processes. The cause of various manifestations on reservoir heterogeneity is different studied scales. This will be discussed in Sect. 1.5.

It should be mentioned that heterogeneity and homogeneity are two ends of a continuous spectrum. This means that the heterogeneity of two systems can be compared and the systems may be more or less heterogeneous. Increasing heterogeneity means increase in the random mixing of the interested parameter (Fitch et al. 2015). So, there are hierarchies in heterogeneity of a reservoir. Key classes and ranges are defined to break these continuous scales and enable us to compare various reservoirs from this point of view. Examples are rock types, hydraulic flow units (HFU), sedimentary facies and even a system tract of a sequence.

While heterogeneity mainly refers to the change in properties in space, anisotropy defines the dependence on direction. In other words, the property is homogeneous if it is independent of position and is isotropic if it is independent of direction. From statistical point of view, when a body is competently homogeneous, subsamples have the same mean values of the property of interest. A good and well-known example in reservoir studies is permeability. This property is heterogeneous in many cases, especially in carbonate reservoirs, but its variations strongly depend on direction. Horizontal permeability is higher than vertical permeability in many cases because particles are deposited perpendicular to their maximum projection area (MPA). The MPA is the maximum projected surface of a particle onto an arbitrary plane. Also, many barriers within reservoir layers such as stylolites and solution seams as well as shale and anhydrite layers are usually horizontal. Stylolites and solution seams are commonly formed due to the dominant overburden pressure, while shale and anhydrite layers are deposited horizontally. In the case of isotropic properties such as permeability, the most exact measurements are made when the character is measured in all directions. It is obvious that this is not possible on core plugs and so just maximum (horizontal) and minimum (vertical) permeabilities are measured in many cases. Commonly, many properties of a petroleum reservoir are anisotropic and heterogeneous. An illustration of the difference between these two concepts can be seen from Fig. 1.1.

In reservoir evaluation studies, many authors refer to reservoir heterogeneity without a proper and exact definition of this term. So, the clear concept of this word is not clear in many cases.

Heterogeneity is clearly defined just when a scale is attributed to this term. In other words, a rock may be homogeneous in one scale and heterogeneous in another one. The scale is variable from micro for studying the heterogeneity of a thin section or a plug to meso-, macro- and gigascale for layers, field and basin, respectively. So, change in scale may create homogeneity out of heterogeneity (Dutilleul 1993). Conversely, integrated homogeneous bodies may create a heterogeneous volume. For example, structures such as layering create heterogeneity in well scale but rock volumes within these layers may be homogeneous and layering has no strong effect on this scale. It also depends on the sample location. If the analyst looks at the layers' contact, the studied volume is also heterogeneous in microscale. In reservoir evaluation, the properties of interest should have a significant impact on fluid storage capacity and fluid flow.

Heterogeneity could be considered both qualitatively and quantitatively. The first strongly depends on analyst's expectations and experiences. It also depends on the scale of the studied volume, as discussed before. User-dependent measurements are always very variable. So, the heterogeneity should be quantified in order to make a direct comparison between two systems (reservoirs). Many statistical parameters have been introduced in the literature. They include coefficient of determination (R^2), coefficient of variation (CV), correlation coefficient (CC), Dykstra–Parsons

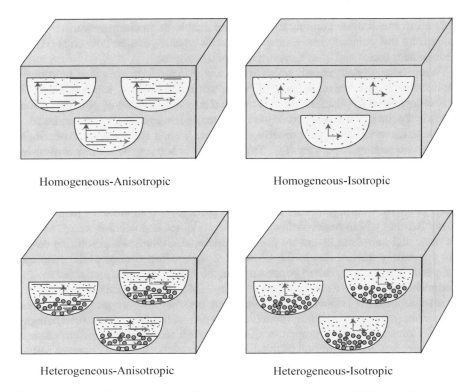

Homogeneous-Anisotropic   Homogeneous-Isotropic

Heterogeneous-Anisotropic   Heterogeneous-Isotropic

Fig. 1.1 A schematic representation of heterogeneity versus anisotropy in tidal channels

coefficient, Lorenz plot and probability distribution function (PDF). The qualitative analysis can be tested with these parameters. For example, geological rock types (Tavakoli 2018) are determined based on the geologist's experience with respect to the depositional properties (facies) and diagenetic processes. The heterogeneity of the petrophysical properties (such as porosity and permeability) of these rock types can be evaluated using the above-mentioned statistical methods. R^2 between porosity and permeability and CV for many petrophysical variables within each rock type are some examples (Nazemi et al. 2018 for example).

Quantifying reservoir heterogeneity is not a simple process. First of all, many variables are involved. Any single parameter is inadequate to properly evaluate the variations of all reservoir characteristics in space and time. For example, many textural (such as micrite content, allochems' size, sorting, interparticle space) or diagenetic parameters (such as dolomitization, dissolution, cementation, compaction) can change the reservoir properties. These parameters are strongly heterogeneous in many carbonate reservoirs. So, they should be included in any statistical calculations. Other quantitative parameters such as porosity and permeability should also be involved. So, a unique unit should be defined to solve the problem. Defining such unit also depends on the scale of interest. For example, hydraulic flow units and electrofacies are homogeneous in microscopic scale, while a sequence stratigraphic unit is defined in field scale. This is discussed in Chaps. 2–4. The second problem is limitations in the number of samples. In many cases, there are not enough samples to adequately address the variations in reservoir properties. The wells are generally about 8 inch (20 cm) in diameter in reservoir part. This is just a small part of the reservoir body. If we consider 1 km distance between two wells, the interwell space is about 5000 times larger than this volume of data! In fact, we are creating this volume just based on such limited data. There is a very interesting scientific story behind this creation. This story is the subject of this book.

Many works have been focused on overcoming this challenge. In fact, we are managing the reservoir heterogeneity in many parts of a carbonate reservoir characterization. Some examples are defining depositional and diagenetic facies, reservoir rock typing, flow unit determination, upscaling, reservoir zonation and sequence stratigraphy.

1.1.1 Related Scientific Definitions

Evaluating the heterogeneity is a very important process in any carbonate reservoir characterization. So, many researchers have attempted to introduce a scientific procedure to define a homogeneous unit and classify the reservoir samples based on their heterogeneity. Readers are familiar with the results of some of these studies. For example, a microfacies defines carbonate samples with specific sedimentological and paleontological characteristics (Flugel 2010). It is obvious that this is a known method to classify depositional properties of carbonate rocks based on their environmental settings. The goal of such classification is interpreting the primary depositional conditions. Regarding the heterogeneity of a carbonate reservoir, each sample has its specific characteristics in petrographical studies and so there will be numerous names for them. Naming the samples based on Dunham (1962) classification combined with Folk (1959) prefixes for the allochems with more than 10% frequency is a good example. For grainstone samples with various amounts of bioclasts and ooids, four different names could be present including ooid grainstone, bioclast grainstone, ooid bioclast grainstone and bioclast ooid grainstone. The presence of another allochem such as intraclast, pellet, peloid or oncoid increases this diversity. Classifying these samples into various microfacies solves the problem. This example can be named bioclast/ooid grainstone and shows deposition in high-energy shola environment. Another familiar example is rock typing. Rock types are samples with similar reservoir properties (Tiab and Donaldson 2015; Tavakoli 2018). There are various methods for determining the rock types of a reservoir, from qualitative geological point of view to quantitative approaches such as RQI/FZI, Lorenz and Winland method.

Some others have limited application in reservoir characterization. For example, Poeter and Gaylord (1990) defined the term "hydrofacies" as a three-dimensional aquifer element in the study of Hanford site aquifer. They believed that if hydraulic properties-related data combine with connection related characteristics, the result will be a hydrofacies. They stated that such definition significantly improves the accuracy of the numerical models. The term used later in both aquifers and petroleum reservoirs for evaluating and managing the heterogeneity (such as Eaton 2006; Engdahl et al. 2010; Bakshevskaya and Pozdnyakov 2013; Finkel et al. 2016; Hsieh et al. 2017; Bianchi and Pedretti 2018; Song et al. 2019). They introduced a new term for an old concept. In fact, the rock type has the same meaning in formation evaluation.

Another example is representative elementary volume (REV) defined by Bear (1972) for the first time. He defined the term "point" against the continuum concept of heterogeneity. Representative value of the property in the point is calculated by simple arithmetic averaging in a volume around this point. It is obvious that this volume should be small enough to be the building block of that body; otherwise, the average value is not a good representative of the property of interest. It should also be larger than just "one" value for a meaningful statistical averaging. A heterogeneous body is assumed. The volume ΔU_i is assumed within this body, containing several values of the property of interest. Pore type in a carbonate reservoir is a good example. The property is moldic porosity in this example. The volume ΔU_i contains several moldic pores. It is obvious that this volume also contains other pore types due to the heterogeneity of the reservoir (Fig. 1.2). The ratio of desired property volume $(\Delta U_v)_i$ to ΔU_i (Eq. 1.1) is defined as (Bear 1972):

$$n_i(\Delta U_i) = (\Delta U_v)_i / \Delta U_i \qquad (1.1)$$

This ratio is always less than or equal to 1 because the largest volume of the property of interest is the same as the assumed volume (ΔU_i). By gradually shrinking

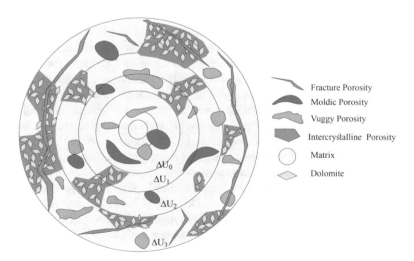

Fig. 1.2 Schematic representation of REV with the pore type example in a carbonate rock sample under the microscope

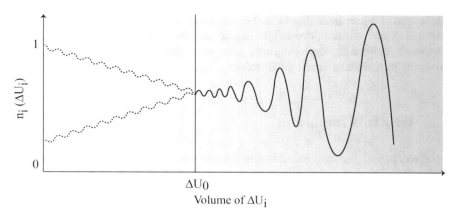

Fig. 1.3 Quantitative definition of REV and the suitable size of a homogeneous sample

the size of ΔU_i, the fluctuations of ΔU_i versus $n_i(\Delta U_i)$ decrease and a nearly flat plateau is formed, because theoretically there is a certain size for ΔU_i where the volume is almost homogeneous (ΔU_0). This size is scale-dependent and can be defined for any property (Fig. 1.3). The volume of ΔU_i where the $n_i(\Delta U_i)$ is nearly a constant value is REV of that property. After this limit, the ratio approaches zero or 1 depends on the position of the property in the body. If the property is in the center of the volume, the smallest volume will be completely filled and the value is 1. In contrast, if the volume of the property inside the smallest ΔU_i is zero, the ratio will be zero, too. As REV is a nearly homogeneous volume, a statistical average can be assigned to its center.

Many researchers have attempted to evaluate the potential of this method in reservoir studies. Bachmat and Bear (1987) explained the concept from mathematical point of view. Brown and his co-workers (2000) used REV for evaluation of the suitable core sample size. Eaton (2006) explained that the REV is the smallest volume with the same governing equations of flow. Nordahl and Ringrose (2008) applied the method to permeability data in heterolithic deposits. Vik and his co-workers (2013) used the method for understanding the proper sample size to characterize a vuggy carbonate reservoir.

The REV is the smallest volume of a reservoir which average values of the properties can be performed. So, it is very useful for selecting the smallest block for a numerical reservoir modeling. It is obvious that primary depositional characteristics as well as diagenetic impacts on the reservoir control the REV. It also depends on the scale. In carbonate reservoirs, facies control the properties in large scales, while diagenesis changes the reservoir characteristics in small scales.

Is it really necessary to define a new term for a homogeneous unit within a reservoir? Regarding the various present terms such as depositional facies, diagenetic facies, pore types, pore facies, rock types, flow units and similar words, it seems that no new term is needed for defining such unit in a reservoir. In fact, terms such as rock type include all properties that should be involved in a reservoir evaluation.

All reservoir parameters should be homogeneous in one rock type and should be different in various ones (Tavakoli 2018). So, a new definition will not be useful. The important problem is developing new procedures for better differentiation of rock groups with similar reservoir performance.

1.2 Why It Is Important?

Reservoir heterogeneity is one of the most important issues in the context of reservoir evaluation. The main goal of static reservoir characterization is determining the volume of hydrocarbon in place (HIP). To calculate HIP, understanding of several parameters is necessary. They include the geometry of the reservoir, the volume of empty spaces (that are not filled with rock matrix) and the portion of these spaces that are filled with hydrocarbon. The architecture of the sedimentary environment is used for determining the reservoir geometry. The concept of facies models (Walker 1984) is used to reconstruct the reservoir geometry and exact correlation of sedimentary units in a 3D space. A facies model is a typical integration of individual facies in a particular sedimentary environment (Walker 1984).

The space available for hydrocarbon storage in a reservoir is explained by the porosity term. It is the ratio of the volume of pores to the bulk volume of the sample and can be achieved using core or wire-line log data. A part of this space is filled by hydrocarbon, which is called hydrocarbon saturation (S_h), and the remaining volume is occupied by water. This part is called water saturation (S_w). The water saturation is the ratio of water to all fluids of the rock and is determined routinely based on the Archie's equations. Ultimately, the HIP is defined as follows (Eq. 1.2):

$$\text{HIP} = V_r \times \Phi \times S_h \tag{1.2}$$

where V_r is the total volume of rock and Φ is porosity. Generally, available data are just a tiny fraction of the total volume of the reservoir. The drill hole diameter is just about 20 cm in reservoir parts. The depth of investigation of most logs is less than 0.5 m. The main problem is distributing and predicting these parameters in interwell spaces. Sometimes, hundreds of meters should be predicted based on such limited data. Each small mistake resulted in a large error in calculating HIP and will change the development plan of the reservoir. Such prediction strongly depends on the knowledge of reservoir heterogeneity and spatial distribution of these properties. Any parameter that controls these changes including rock texture and diagenetic impact is also included. The permeability should also be considered to correct prediction of the production rate and to decide whether the field should be developed or not. There are many specialized software with specific mathematical algorithms for predicting reservoir parameters in undrilled spaces. All of them follow specific rules which were developed based on the concept of heterogeneity.

Heterogeneity also plays an important role in sample selection for different tests. After completion of routine and geological core analysis, some samples are selected for the special core analysis (SCAL). These experiments are expensive and time-consuming (McPhee et al. 2015). So, sample selection is very important for these tests. Limited number of samples are tested, and the results are assigned to the overall volume of the reservoir. In fact, these samples should be representative of all other samples in the well. In a heterogeneous reservoir, samples are divided into approximately homogeneous groups based on geological and routine core analysis (RCAL) data. SCAL samples are selected from these groups considering the reservoir quality of each category. It is obvious that no sample is selected from non-reservoir group (or may be groups). The best results are achieved when the heterogeneity within each group is minimum, and so the results can be generalized to the other samples within this cluster. Routinely, the relationship between two parameters is derived as an equation. Changes in porosity and permeability with increasing overburden pressure are a good example. Some samples are selected from each group for overburden tests, and an equation is obtained by correlating the changes in porosity or permeability with overburden pressure. This equation is used for converting ambient porosity and permeability of other samples in this group to the properties in overburden reservoir pressure.

In larger scales, heterogeneity is important in identifying reservoir boundaries. Estimating the HIP and the determining the location of new wells strongly depend on these boundaries. It is also important in well productivity and optimizing the oil and gas recovery from the reservoir. Unexpected decline in production rate may be the result of strong heterogeneity in a reservoir. Such changes in productivity have strong effects on financial investment in the field. A comprehensive knowledge of reservoir heterogeneity is also necessary during secondary, third and fourth recovery. Water and gas injection would be useful when they are injected into the appropriate well with predictable pattern of fluid distribution. The heterogeneity also may cause unexpected migration and water production from hydrocarbon-producing wells. The geomechanical properties of the rocks also vary as a function of reservoir properties. These changes are crucial for drilling new wells.

Evaluating reservoir heterogeneity is not limited to fluid-related properties. It is also important to understand primary depositional settings and secondary diagenetic processes. These geological characters, in turn, are used for reconstructing the horizontal and vertical distribution of facies and sedimentary environments, sea-level changes and dominant diagenetic environments in each time and location in the reservoir. This information enables the construction of more accurate 2D maps and 3D models.

1.3   Sources of Heterogeneity

Rock properties of a reservoir are the result of primary environmental factors and/or secondary diagenetic processes. In a carbonate reservoir, the first one is affected by physical, chemical and biological conditions of the depositional settings. Among them, the energy of the environment has a decisive role and specifies the final textures of the rocks. Grainstones are formed in high-energy depositional settings. In the lower energy environments, more mud and less allochems are deposited. The grainstone facies gradually changes to packstone, wackestone and mudstone with decreasing environmental energy. Allochems also change from high-energy ooids to low-energy pellets. The faunal diversity and frequency vary as a function of environmental conditions. A little change in these conditions leads to the change in their frequency, form, type and size. So, different textures and allochems are present in different parts of the basin. Terrigenous input from the land also changes the lithology of the rocks. In carbonate environments, their frequency is less than 50%. Beyond this limit, the samples are not classified as carbonate and the environment change to a siliciclastic dominated setting. These changes are more obvious in rimmed shelves. The facies change rapidly from intertidal evaporites and low-energy mudstones to lagoonal wackestones and packstones. The high-energy oolithic or bioclastic shoals separate the slope from lagoon. The facies and environmental energy are different in various parts of the slope. The depositional facies are more homogeneous in deep basin environment because the energy is nearly constant in a wide area. Usually, the basin part of a carbonate depositional setting has no reservoir potential.

Changes in carbonate textures are more gradual in ramps. The low-angle substrate causes lower heterogeneity of the facies because physical, chemical and biological conditions change gradually across the depositional settings.

While water depth changes at various parts of a carbonate depositional environment and causes facies heterogeneity in the rocks with the same age, sea-level fluctuations affect the conditions in vertical succession. With increasing or decreasing water depth due to the sea-level changes, different samples with distinct textures and allochems are deposited. In a carbonate–evaporite environment, the evaporative minerals such as gypsum are deposited and form the compartmentalizer layers.

Although facies change is the main reason for heterogeneity in a carbonate reservoir, its variations are routinely observed in meter to tens of meter scale. In contrast, diagenetic processes modify the rock properties at smaller scales. These processes start just after deposition. Marine isopachous rim cements, micritization, dolomitization, anhydrite cementation, precipitation of calcite blocky cements, fracturing, chemical and mechanical compaction and dissolution are some important diagenetic processes in a carbonate reservoir. Diagenetic processes take place in marine, meteoric and burial realms with different effects on various parts of a facies. While aragonitic ooids are easily dissolved in meteoric diagenetic environment, interparticle low-magnesium calcite cements remain unaltered. Different mineralogy of various shells also affects diagenetic processes and the heterogeneity of the reservoir (Fig. 1.4).

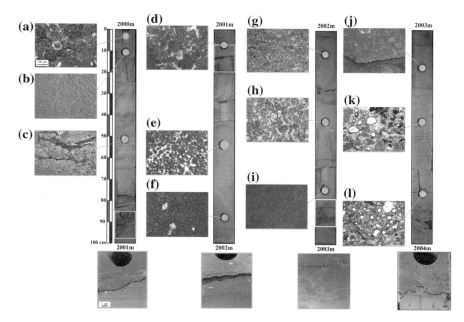

Fig. 1.4 Facies and diagenetic heterogeneity in a Permian–Triassic carbonate core from Persian Gulf area. Alternation of grainstone (**a, d, e, h, k, l**) and mudstone (**b, c, f, i, j**) within 4 meters of cores is observable. A packstone (**g**) is also present. Diagenetic changes are at microscopic scale. Different sizes of dolomite crystals in one thin section (**a, b, c**), different micritization processes on one sample (**d**) and different rates of dissolution on grains (**k, l**) are obvious. The scales of all thin section images are the same as subfigure "**a**," plane-polarized light

Jung and Aigner (2012) introduced six factors controlling carbonate geobodies and so their reservoir quality. They include geological age (depo-time), type of carbonate platform (depo-system), facies belts (depo-zone), 3D geometrical classification (depo-shape), building blocks of the depo-shape (depo-element) and carbonate lithofacies (depo-facies). Example of a depo-system is a carbonate ramp or rimmed shelf. Depo-zones are subenvironments such as intertidal or lagoon. Bars or mounds are examples of depo-shape, and their various subenvironments such as flank or crest are depo-elements. Finally, rock textures and allochems make the lithofacies. Diagenetic processes must be included which have been ignored in this study. These processes usually follow the general trend of these factors, especially in the case of early diagenesis.

1.4 Materials

Which data sources are used for understanding various aspects of heterogeneity in a carbonate reservoir? Generally, two types of data are available from a reservoir including direct and indirect data. Direct data include information derived from cores and cuttings. Various data types such as geological (Tavakoli 2018), petrophysical (McPhee et al. 2015), geomechanical (Zoback 2007) and geochemical (Ramkumar 2015) are derived from cores. Cores are the most reliable source of reservoir information. In addition, providing some types of data from other sources is not possible. Permeability and rock texture are some of these examples. Lithology, type and frequency of allochems, microscopic sedimentary features such as bioturbation, presence of opaque minerals, laminations, mud cracks, brecciation and fenestral fabric of the samples, type and frequency of various pore types, fractures, cements and compaction features are retrieved from thin section analysis. Macroscopic features such as fractures, facies changes, large intraclasts, various types of surfaces, sedimentary structures and large pores are recorded in macroscopic core description. Computed tomography scanning (CT scan) of the cores is used for identifying fractures and also calculating some petrophysical properties. Various petrophysical properties are derived from cores in routine and special phases of a core analysis project. High-resolution sampling yields more information about the heterogeneity of the property of interest. Anyway, in a standard core analysis project, four samples are taken from 1 m of core (Tavakoli 2018).

Cuttings are another source of direct information, but their application is limited due to the small sample sizes. Thin sections can be prepared from cuttings, but pore types could not be distinguished. Running most of the routine tests is not possible on cuttings. During drilling, rocks continuously fall out and mix with the bottom hole samples.

Wire-line logs and seismic data provide indirect measurements from rocks and fluids of the reservoir. Logs provide different types of information such as natural gamma radiation, density, porosity, sonic velocity and electrical resistivity. Petrophysical analysis based on these data yields important information from the reservoir. The routine sampling interval is 0.1524 m for these data. In contrast, seismic sections provide information at larger scales. They are used for macroscopic to megascopic reservoir studies. Geometry of the reservoir, large structural features such as main faults, traps and contacts are some examples (Fig. 1.5).

1.5   Heterogeneity Scales

The heterogeneity concept is always accompanied by a specific scale. In fact, a sample may be homogeneous at one scale and heterogeneous at another. A core plug may be homogeneous with respect to its pore types, while the core itself is heterogeneous. Heterogeneity hierarchies also depend on the property. While facies may be homogeneous in 1 m of core, they vary considerably in well scale. Heterogeneity and homogeneity are end-members of a continuous spectrum. So, key units should be defined to classify these hierarchies. Samples have different sizes, depending on the objectives of the studies. The REV which was discussed in Sect. 1.1.1 is a

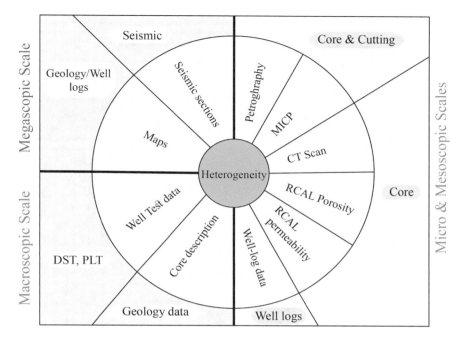

Fig. 1.5 Various data types which are used for heterogeneity analysis of a carbonate reservoir. The heterogeneity scale is also important, and so some of these classifications may change according to the scale

mathematical representation of how the sample size plays an important role in its heterogeneity.

Heterogeneity exists at micro-, meso-, macro-, mega- and gigascales in a reservoir. Microscopic heterogeneities include facies characteristics, diagenetic effects, pore types, pore throat sizes, grain shape, size and packing as well as mineralogy of the samples. These heterogeneities are measured using thin sections, scanning electron microscope (SEM), mercury injection capillary pressure (MICP) tests, core CT scanning and wire-line logs. Porosity–permeability relationships are presented, various classification schemes (such as pore typing) are applied to data, and electrofacies (EF) are defined from wire-line logs to overcome the heterogeneity challenges at this scale. At mesoscopic scale, fine-scale data are integrated at well scale to construct traceable homogeneous units in the reservoir. Defining sedimentary environments and hydraulic flow units (HFUs), stratigraphic correlations and upscaling are some examples. Macroscopic heterogeneities include field-scale variables such as stratification, compartmentalization, sequence stratigraphy, reservoir zonation and lateral property trends (Fig. 1.6). The megascopic scale of heterogeneity represents the basin-scale parameters such as tectonic activities and the role of structural features as well as widespread depositional conditions. So, various data are combined to understand the overall heterogeneity of a reservoir (Fig. 1.7).

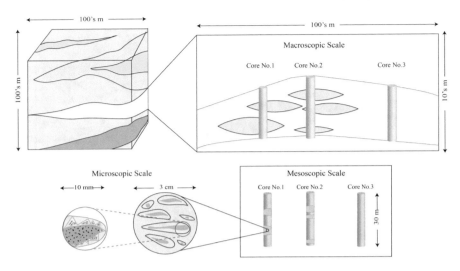

Fig. 1.6 Various scales from microscopic to field heterogeneity

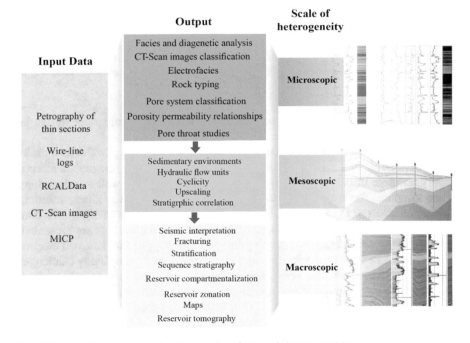

Fig. 1.7 Input data, outputs and various scales of reservoir heterogeneities

The sampling, tracking, correlation and mapping of strata are not limited in surface outcrops, but the resolution of the tools in subsurface studies is a major problem. Direct observation is not possible in subsurface, and so the sampling rate and scale of the analysis strongly depend on the tool resolution.

References

Amaefule JO, Altunbay M, Tiab D, Kersey DG, Keelan DK (1993) Enhanced Reservoir Description: using core and log data to identify hydraulic (flow) units and predict permeability in uncored Intervals/Wells. In: 68th annual technical conference and exhibition. Houston, TX, Paper SPE26435

Bachmat Y, Bear J (1987) On the concept and size of a representative elementary volume (REV). In: Bear J, Corapcioglu MY (eds) Advances in transport phenomena in porous media. NATO Advanced Study Institute on Fundamentals of Transport Phenomena in Porous Media Series E, vol 128. Kluwer Academic Publishers, Dordrecht, Boston, MA, pp 3–19

Bakshevskaya VA, Pozdnyakov SP (2013) Methods of modeling hydraulic heterogeneity of sedimentary formations. Water Resour 40(7):767–775

Bear J (1972) Dynamics of fluids in porous media. American Elsevier Publishing Co, New York

Bianchi M, Pedretti D (2018) An entrogram-based approach to describe spatial heterogeneity with applications to solute transport in porous media. Water Resour Res 54(7):4432–4448

Brown GO, Hsieh HT, Lucero DA (2000) Evaluation of laboratory dolomite core sample size using representative elementary volume concepts. Water Resour Res 36(5):1199–1207

Dunham RJ (1962) Classification of carbonate rocks according to depositional texture. In: Ham WE (ed) Classification of carbonate rocks. AAPG Memoir 1, Oklahoma

Dutilleul P (1993) Spatial heterogeneity and the design of ecological field experiments. Ecology 74:1646–1658

Eaton TT (2006) On the importance of geological heterogeneity for flow simulation. Sediment Geol 184:187–201

Engdahl NB, Vogler ET, Weissmann GS (2010) Evaluation of aquifer heterogeneity effects on river flow loss using a transition probability framework. Water Resour Res 46:W01506

Finkel M, Grathwohl P, Cirpka OA (2016) A travel time-based approach to model kinetic sorption in highly heterogeneous porous media via reactive hydrofacies. Water Resour Res 52(12):9390–9411

Fitch P, Lovell MA, Davies SJ, Pritchard T, Harvey PK (2015) An integrated and quantitative approach to petrophysical heterogeneity. Mar Petrol Geol 63:82–96

Flugel E (2010) Microfacies of carbonate rocks, analysis, interpretation and application. Springer, Berlin

Folk RL (1959) Practical petrographic classification of limestones. AAPG Bull 43(1):1–38

Frazer GW, Wulder MA, Niemann KO (2005) Simulation and quantification of the fine-scale spatial pattern and heterogeneity of forest canopy structure: A lacunarity-based method designed for analysis of continuous canopy heights. For Ecol Manag 214(1–3):65–90

Hsieh AI, Allen DM, MacEachern JA (2017) Upscaling permeability for reservoir-scale modeling in bioturbated, heterogeneous tight siliciclastic reservoirs: Lower Cretaceous Viking Formation, Provost Field, Alberta, Canada. Mar Petrol Geol 88:1032–1046

Jung A, Aigner T (2012) Carbonate geobodies: hierarchical classification and database—a new workflow for 3D reservoir modelling. J Petrol Geol 35:49–66

Kolodizie SJ (1980) Analysis of pore throat size and use of the Waxman-Smits equation to determine OOIP in Spindle Field, Colorado. SPE paper 9382 presented at the 1980 SPE Annual Technical Conference and Exhibition, Dallas, Texas

Li H, Reynolds J (1995) On definition and quantification of heterogeneity. Oikos 73:280–284

Lucia FJ (1995) Rock-fabric/petrophysical classification of carbonate pore space for reservoir characterization. Am Assoc Petr Geol B 79(9):1275–1300

Lucia FJ, Conti RD (1987) Rock fabric, permeability, and log relationships in an upward-shoaling, vuggy carbonate sequence. Bureau of Econ Geol Geol Circular 87–5

McPhee C, Reed J, Zubizarreta I (2015) Core analysis: a best practice guide. Elsevier, UK

Nazemi M, Tavakoli V, Rahimpour-Bonab H, Hosseini M, Sharifi-Yazdi M (2018) The effect of carbonate reservoir heterogeneity on Archie's exponents (a and m), an example from Kangan and Dalan gas formations in the central Persian Gulf. J Nat Gas Sci Eng 59:297–308

Nordahl K, Ringrose PS (2008) Identifying the representative elementary volume for permeability in heterolithic deposits using numerical rock models. Math Geol 40(7):753–771

Poeter EP, Gaylord DR (1990) Influence of aquifer heterogeneity on contaminant transport at the Hanford Site. Ground Water 28(6):900–909

Ramkumar Mu (ed) (2015) Chemostratigraphy: concepts, techniques and application. Elsevier, Amsterdam

Song X, Chen X, Ye M, Dai Z, Hammond G, Zachara JM (2019) Delineating facies spatial distribution by integrating ensemble data assimilation and indicator geostatistics with level-set transformation. Water Resour Res 55:2652–2671

Tavakoli V (2018) Geological core analysis: application to reservoir characterization. Springer, Cham, Switzerland

Tiab D, Donaldson EC (2015) Petrophysics, theory and practice of measuring reservoir rock and fluid transport properties. Gulf Professional Publishing, Houston

Vik B, Bastesen E, Skauge A (2013) Evaluation of representative elementary volume for a vuggy carbonate rock—part: porosity, permeability, and dispersivity. J Petrol Sci Eng 112:36–47

Walker RG (1984) General introduction: facies, facies sequences and facies models. In: Walker RG (ed) Facies models, 2nd edn. Geological Association of Canada, Geoscience Canada Reprint Series 1

Weber KJ (1986) How heterogeneity affects oil recovery. In: Lake LW, Carroll HB (eds) Reservoir characterization. Academic Press, New York, pp 487–544

Zhengquan W, Qingeheng W, Yandong Z (1997) Quantification of spatial heterogeneity in old growth forests of Korean Pine. J For Res 8:65–69

Zoback MD (2007) Reservoir Geomechanics. University Press, Cambridge, England

Chapter 2
Microscopic Heterogeneity

Abstract Heterogeneity is variation in space and time which strongly depends on the scale of the study. This chapter focuses on microscale heterogeneities in a carbonate reservoir. At first, available data and methods are discussed to understand the microscopic heterogeneity in carbonates. Then, various methods are introduced to overcome this challenge. All reservoir properties depend on geological attributes, and so facies analysis and diagenetic impacts on the carbonate reservoirs are discussed. Facies grouping and diagenetic facies are introduced to classify different geological properties into similar categorizes. CT scan images are used for classifying the samples. Intelligent systems are widely used for this purpose. Porosity–permeability relationships are one of the most important criteria which are used for understanding the homogeneity of a unit. This is different for limy and dolomitic reservoirs. Interparticle, moldic and vuggy porosities with completely different petrophysical behaviors are dominant in limy carbonate reservoirs. Intercrystalline porosities between dolomite crystals generally increase the homogeneity of a sample, but the result also depends on the type and amount of dolomitization. In contrast to limy reservoirs, dolostones have small but uniform pore throat size distribution. Pore system classifications, rock typing methods and electrofacies are used for managing the heterogeneity in microscale. Pore types significantly control the petrophysical behavior of the carbonate reservoirs. They have been divided into various groups from different points of view. Rock typing procedure starts with the study of geological properties and continues with the incorporation of the petrophysical characteristics of the samples. Wire-line log data are grouped based on their similar readings and then are correlated with the previously determined rock types. The final unit is used for predicting reservoir properties in 3D space between the wells.

2.1 Facies Analysis

Facies is a set of properties which characterize a sedimentary rock sample. The term introduced by Gressly (1838) for the first time and has been the subject of considerable debate so far. Recently, the term microfacies was defined by Flugel (2010) as all characters of a carbonate sample from thin section properties to hand

© The Author(s), under exclusive license to Springer Nature Switzerland AG 2020   17
V. Tavakoli, *Carbonate Reservoir Heterogeneity*,
SpringerBriefs in Petroleum Geoscience & Engineering,
https://doi.org/10.1007/978-3-030-34773-4_2

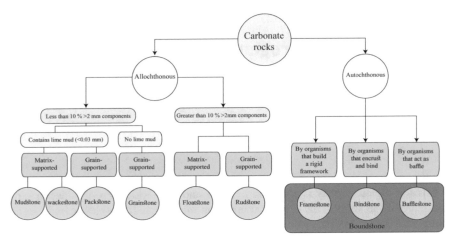

Fig. 2.1 Carbonate classification by Dunham (1962) modified by Embry and Klovan (1971)

specimen observations. In fact, it is equivalent to petrofacies plus lithofacies in siliciclastic rocks. Routinely, the term facies is used instead of microfacies. The terms
facies and microfacies are used synonymously in this book. Dunham (1962) classification is commonly used in combination with the most frequent allochem or two
allochems (more than 10%) for facies naming in carbonates (Fig. 2.1). Such classification reflects the environmental energy as well as frequent allochems and so
represents the sedimentary environment. Facies have been deposited in the same
depositional conditions, and so they should also contain similar reservoir properties.
Following deposition, rock characteristics are affected by diagenesis. So, both primary conditions and secondary changes determine the final reservoir quality of a
formation.

Using the above definition, there will be a lot of facies in a heterogeneous carbonate formation. So, facies are classified into facies groups. The main criteria for
defining a facies group are sedimentary environment. The main purpose of facies definition is representing its environmental conditions, and so similar facies are grouped
into facies groups (Tavakoli 2018). This is independent of their reservoir characteristics, but as mentioned, with respect to their similar genetic conditions, the reservoir
properties should also be the same. Therefore, petrophysical behavior of the samples
can be predicted using facies distribution in a carbonate reservoir.

Facies heterogeneity is a direct result of change in environmental conditions,
especially energy of the depositional settings. They may vary in decimeters to tens
of meters. Figure 2.2 shows carbonate textures in two Iranian formations including
Dalan (Late Permian) and Dariyan (Aptian) from one offshore well in Persian Gulf.
It can be seen in this figure that facies vary considerably in Dalan Formation, while
in Dariyan, facies are constant in nearly tens of meters. It is related to the depth
of deposition. While Dalan Formation has been mainly deposited in inner ramp
setting with water agitation, a mid-to-inner ramp depositional setting is proposed

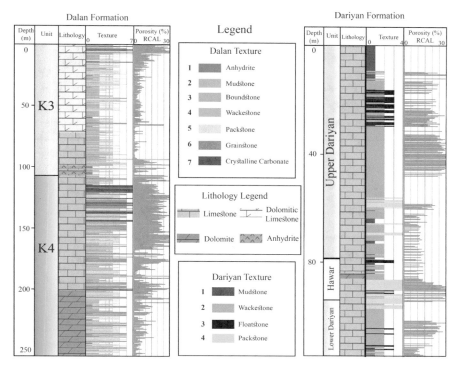

Fig. 2.2 Comparison of facies heterogeneity of two Iranian Dalan (Late Permian) and Dariyan (Aptian) formations. Lithology and porosity of the wells are obvious (modified from Tavakoli and Jamalian 2018)

for Dariyan Formation (Tavakoli and Jamalian 2018). So, this formation is mainly composed of mud-dominated samples. Lithology as well as porosity distribution in both formations is obvious (Fig. 2.2).

Many researchers have tried to find a relationship between facies distribution and reservoir properties. Lucia (1995) tried to relate grain size and so sample texture to porosity and permeability of the rocks. He defined three petrophysical classes; each class is related to a special rock fabric and porosity–permeability distribution. Lucia's works have been used by many authors to evaluate the relationship between porosity–permeability and rock textures in various carbonate reservoirs of the world. Plots of porosity versus permeability in four Iranian carbonate reservoirs (Fig. 2.3) show that grain-dominated samples generally have higher porosity and permeability. Grainstones and packstones have higher porosity and permeability compared to mudstones and wackestones. The formations include Kangan and Dalan equivalent to Khuff reservoir, the largest gas reservoirs of the world, Ilam and Sarvak, two Iranian oil-bearing formations.

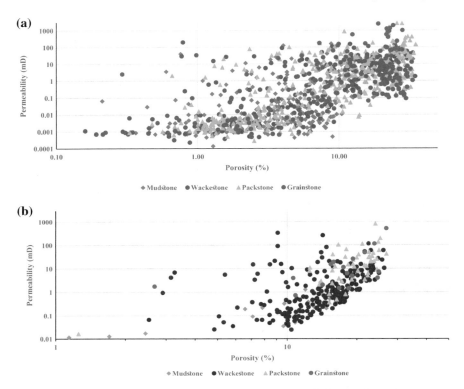

Fig. 2.3 Plot of porosity versus permeability for different textures in four Iranian carbonate reservoirs including Permian–Triassic Dalan and Kangan in offshore Persian Gulf (**a**) and Cretaceous Ilam and Sarvak in southwest of Iran (**b**)

2.2 Diagenetic Impacts

Sedimentary rocks are affected by diagenetic processes immediately after deposition. The heterogeneity of these rocks changes due to diagenesis in marine, meteoric and burial realms. Cementation, dissolution, compaction, neomorphism, dolomitization and fracturing are routine diagenetic impacts in carbonate reservoirs. Some of these processes enhance reservoir performance and increase porosity and permeability, while others reduce the reservoir quality of the carbonate formations. Facies distribution follows the general trend of sedimentary environments, while diagenetic processes follow the facies characteristics in many cases, especially in early diagenesis (Fig. 2.4). The aragonitic ooids are more prone to dissolution in meteoric environments. Widespread dolomitization occurs in seepage–reflux dolomitization model which happens in peritidal environments. This process is fabric-selective in many cases. So, the scale of changes is smaller than facies. Dolomitization is one of the most important diagenetic processes in carbonate reservoirs and has major effects on reservoir properties. So, limestone and dolomite samples are considered separately.

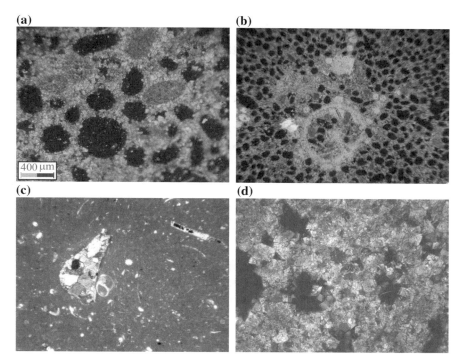

Fig. 2.4 Selective diagenesis in three carbonate reservoirs with different ages. Selective dissolution of ooids in Arab (Cretaceous) Formation (**a**), dissolution of peloids and bioclasts and anhydrite cementation in empty spaces of dissolved bioclasts in Arab Formation (**b**), selective dissolution and calcite drusy cementation of bioclasts in Dariyan (Albian) Formation (**c**), dissolution of allochems and matrix dolomitization in Oligo–Miocene Asmari Formation (**d**). All samples are from oil reservoirs which are located in southwest Iran. Photographs have been taken under the cross-polarized light with the same scale

2.2.1 Limestones

Calcite is the main constituent mineral in limestone reservoirs. It could be precipitated as aragonite, low-magnesium calcite (LMC) or high-magnesium calcite (HMC). Aragonite and HMC are the metastable phases of calcite and are gradually transformed to more stable LMC. The calcareous fauna also have different shells' mineralogy and so increase the heterogeneity of the carbonate formations. So, limestone reservoirs are commonly heterogeneous with respect to their mineralogy. Such difference strongly affects their later stages of diagenesis. Aragonite and low-Mg calcite are more susceptible to diagenetic processes such as dissolution and dolomitization. This mineral-selective diagenesis changes the heterogeneity and reservoir quality of the formations. Dissolution of aragonitic ooids creates the separate molds. High porosity and low permeability are the main characteristics of such reservoirs. Dissolved minerals are precipitated as isopachous, blocky, bladed or drusy cements. Calcite cementation also changes the heterogeneity of the carbonate samples at microscopic scales. Isopachous cements retain the original framework of the samples. They build a solid structure by binding the allochems in grain-dominated samples. Such cementation prevents later compaction and so maintains the facies homogeneity of the samples. In contrast, blocky cements are precipitated in larger pores (Ehrenberg and Walderhaug 2015), and so this type of cementation increases the heterogeneity of the carbonate formations. Micrite particles are subjected to aggrading neomorphism. Original fabrics of carbonate allochems are altered to micrite in marine environments. It may affect the whole allochem or just its crust and make micrite envelope around the grains. This envelope prevents more diagenetic impacts (such as compaction) on allochems. Lamination is not a routine structure in carbonate rocks, but it is present in some cases. For example, stromatolites have original laminated structures. They generally live in peritidal environments but can also develop to lower intertidal or even other shallow environments (Tavakoli et al. 2018). Deposition of micrite particles in calm lagoonal conditions also creates fine laminae in these deposits. In such laminations, geological and petrophysical properties are nearly the same in horizontal direction but they are strongly different in vertical column.

Bioturbations also increase heterogeneity in carbonate strata. This process can also change the reservoir properties. Fenestral fabrics are mainly arranged concordant to stratification and increase the porosity of the samples if they are not filled with cements. This type of porosity has no strong effect on reservoir development. The effects of some other factors such as mud cracks, opaque minerals and brecciation are also negligible.

Routinely, there are various pore types in a carbonates formation. The primary interparticle porosities are uniformly distributed in the rocks. They are formed primarily in high-energy depositional settings and are seen in grain-dominated textures. In contrast, distribution of secondary porosities such as moldic and vuggy is not uniform. They generally affect the rocks according to their original mineralogy which is different for various allochems. Fractures crosscut the cements, matrix and also allochems. They increase heterogeneity of the formations especially at the time of production. The fracturing process and its effect on carbonate reservoir heterogeneity are discussed in Sect. 4.2.

Physical compaction affects the rocks in greater depths of burial. The result of such compaction is porosity reduction, broken grains, concave–convex contacts and bending of ductile grains. Chemical compaction-related structures have strong effects on reservoir properties (e.g., Bruna et al. 2019). Their formation, frequency and orientation strongly depend on principal stress direction of the area. This is lithostatic pressure in many cases, and so they are parallel to lamination. Generally, stylolites and solution seams are flow barriers and so reduce permeability (e.g., Mehrabi et al. 2016). Stylolites also may leak depending on collecting material and the size of spikes (Koehn et al. 2016). So, they change the heterogeneity of the reservoir rocks. The important point is that their past and present-day impacts on reservoir quality of the carbonate formations are different (Bruna et al. 2019).

2.2.2 Dolomitic Reservoirs

Dolomite reservoirs, their dolomitization models and textures have been documented in many studies (Sibley and Gregg 1987; Lucia and Major 1994; Warren 2000; Machel 2004; Carnell and Wilson 2004; Saller 2004; Huang et al. 2014; Tavakoli and Jamalian 2019). Dolostones host about 50% of the world hydrocarbons, and so they are still the subject of many studies. Almost all dolomites are formed in secondary diagenetic environment from precursor calcite in the process of dolomitization. In this process, half of the Ca ions are replaced by Mg in an interlayer structure.

Dolomites are classified based on their shape and size. Friedman (1965) introduced the idiotopic, hypidiotopic and xenotopic terms for dolomite crystals' shape. Sibley and Gregg (1987) classified the boundary crystal shapes as planar-s (subhedral), planar-e (euhedral) and non-planar. In euhedral texture, the intercrystalline area is filled by another mineral or remains empty. So, such texture generally increases the heterogeneity of the rock (Fig. 2.5a). In the case of uniform distribution, the sample has the same heterogeneity with different mineralogy (Fig. 2.5b). In contrast, rocks with subhedral texture with compromise boundaries and many crystal junctions exhibit less heterogeneous (Fig. 2.5c, d). In non-planar dolomites, the sample is completely dolomitized and so heterogeneity reduces to its lowest level (Fig. 2.5e). Such diversities also affect the reservoir quality of the rock. At first, euhedral dolomites are scattered in matrix in most cases and so they have little impact on reservoir properties. With increasing dolomite content, the porosity also increases. So, samples with planar-s textures generally have more porosity (Fig. 2.5c). Reservoir quality decreases again with increasing dolomitization rate and dolomite cementation in the samples. These samples routinely have non-planar xenotopic texture (Fig. 2.5e). Evaporative minerals (especially anhydrite) are present in carbonate reservoirs in most cases. They also reduce reservoir properties and increase heterogeneity of the samples (Fig. 2.5f). It should be noted that different textures increase or decrease the heterogeneity of the sample by changing both lithology and porosity in the reservoir rocks.

With respect to the original fabric of the carbonate reservoir rocks, dolomites may be fabric-selective or fabric-destructive. Fabric-selective dolomitization preserves the original texture of the host rock (Fig. 2.6a). The metastable minerals such as aragonite or low-magnesium calcite are replaced at the early stages of diagenesis. So, some parts of the rock are dolomitized, while others are still calcite (or may be aragonite). Such dolomitization increases the heterogeneity of the rock by changing lithology of various parts of the sample. In contrast, in fabric-destructive dolomitization, the main parts of the rock including matrix, allochems and cements are converted to dolomite. This process resulted in more homogeneity of the sample (Fig. 2.6b) unless various dolomite generations are present. For example, the metastable minerals may be converted to dolomite in seepage–reflux model in early diagenesis. These dolomites are fine crystalline with xenotopic crystals. Dolomitization continues in burial realm, and some large planar-e dolomite crystals are formed. In this case, the sample has been pervasively dolomitized but is still heterogeneous.

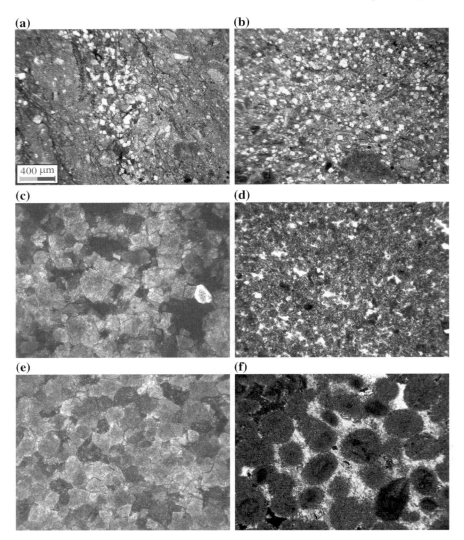

Fig. 2.5 Scatter dolomites in micrite have no strong effect on reservoir quality but increase heterogeneity in microscopic scale (**a**, Upper Cretaceous Tarbur Formation); such disperse dolomite also may have neither effect on heterogeneity nor reservoir properties (**b**, Upper Cretaceous Tarbur Formation), planar-s dolomites with high porosity with large (**c**, Oligo–Miocene Asmari Formation) and fine size (**d**, Upper Permian Dalan Formation) crystals, non-planar (xenotopic) homogeneous dolomites with no porosity (**e**, Oligo–Miocene Asmari Formation) and anhydrite cements which increase heterogeneity and reduce porosity of the dolomitic samples in carbonate–evaporite systems (**f**, Upper Permian Dalan Formation). All views have the same scale. Subsamples **a**, **b** and **d** have been taken under plane-polarized light and the remaining under cross-polarized

(a)   **(b)**

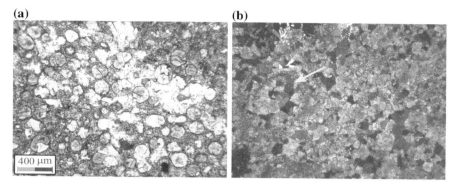

Fig. 2.6 Fabric-selective dolomitization in Early Triassic Kangan Formation (**a**, plane-polarized), fabric-destructive dolomitization in Oligo–Miocene Asmari Formation (**b**, cross-polarized). Ghost of a bioclast is still visible after pervasive dolomitization (arrow). Two views have the same scale

The origin of porosity in dolomitic reservoirs is still controversial. It may be related to dolomitization process or inherited from the precursor limestones (Purser et al. 1994). In mole-for-mole dolomitization up to 13% porosity is generated (Tucker and Wright 1990; Machel 2004). Some researchers (e.g., Saller and Henderson 1998, 2001) believe that in the volume-for-volume replacement, porosity can be constant or even decreases. Combining previous studies, it can be concluded that the porosity of the precursor limestones increases during dolomitization process (Tavakoli and Jamalian 2019). So, some part of the porosity is inherited from primary depositional fabric, while another part is developed during dolomitization. Dolomitization creates intercrystalline porosity or enhances the primary interparticle porosities (Tucker and Wright 1990; Machel 2004; Tavakoli and Jamalian 2019). Both of these pore types are connected, and so permeability also increases during this process. This leads to less heterogeneity in carbonate reservoirs because it improves the porosity–permeability relationships. Both early and late dolomites have the same effect from this point of view. It is obvious that the dolomite fabric is also important in each case.

2.2.3   Diagenetic Facies

Diagenesis has a strong effect on reservoir properties and can decrease or increase the porosity and permeability of the carbonates. So, defining a unit with similar diagenetic characteristics is useful in considering reservoir heterogeneity. From the definition of the term "facies" (Gressly 1838), it can be concluded that diagenetic facies refers to the rocks with the same diagenetic impacts. In fact, such as facies itself, the diagenetic facies is also defined to classify the reservoir rocks based on their diagenetic properties. Zou and his co-workers (2008) believed that with respect to the effect on reservoir properties, there are two types of diagenetic facies including constructive and destructive. These are the result of combined effect of structure, fluids, temperature and pressure on precursor sediments. The rock properties of the samples within these groups routinely have the same depositional characteristics, and so the groups can be used for determining spatial distribution of reservoir parameters. The main data source for classifying diagenetic facies is core and equivalent outcrop sections. Efforts have been made to relate diagenetic processes and impacts to wire-line logs (such as Cui et al. 2017; Lai et al. 2019). The samples can be classified based on diagenetic minerals, processes and environments. The main diagenetic mineral in carbonates is dolomite which has been discussed in Sect. 2.2.2. The main diagenetic processes are micritization, recrystallization, marine cementation, dissolution, dolomitization, compaction (physical and chemical), burial cementation and fracturing. Three diagenetic environments in carbonates include marine, meteoric and burial. Diagenetic processes are discussed within the framework of their environments. In many cases, meteoric and shallow burial diagenesis enhances reservoir properties, while burial processes decrease both porosity and permeability of the samples. Fracturing in late diagenetic environment increases the permeability of the rocks. Burial dissolution is exceptionally effective in porosity and permeability enhancement in carbonate reservoirs.

Reservoir quality is the most important criteria for classifying the samples in a carbonate reservoir. So, a diagenetic facies is routinely defined based on its constructive or destructive effects on reservoir properties. The main constructive processes include dissolution and dolomitization which create moldic and intercrystalline porosity, respectively. Destructive processes are compaction and cementation. The final reservoir quality is the result of both primary facies characteristics as well as secondary diagenetic changes, and so both of them should be considered to define a representative unit. Wang and his co-workers (2017) based on their work in a sandstone reservoir defined the sedimentary–diagenetic facies as a combination of litho- and diagenetic facies. In carbonate reservoirs, a sedimentary–diagenetic facies is defined as a combination of micro- and diagenetic facies. It is more useful in understanding the heterogeneity of a reservoir, as it integrates the effects of both variables. Various primary characteristics as well as diagenetic processes influence a carbonate reservoir. Accordingly, definition of such facies is complicated (Fig. 2.7). In fact, this definition is very close to the concept of geological rock type (GRT) (Tavakoli 2018).

Such as any other homogeneous unit, the diagenetic facies should be investigated at microscale and then expand to meso- and macroscale through macroscopic core analysis, comparison with wire-line log data, upscaling, stratigraphic correlation, mapping and statistical modeling (Fig. 2.8). They also can be used for more accurate prediction of reservoir properties, such as permeability. Each diagenetic facies can have its specific formula with particular coefficients for such predictions.

2.2.4 Quantitative Diagenesis

Quantitative data are required to define a representative unit for understanding reservoir heterogeneity and prediction of reservoir properties in a 3D space. Many

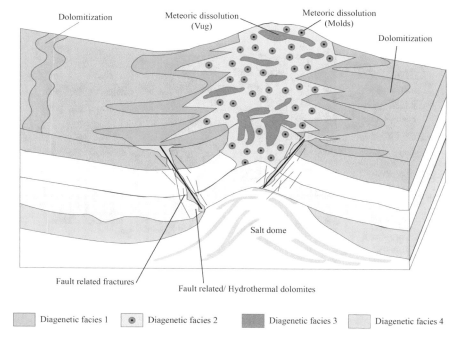

Diagenetic facies 1   Diagenetic facies 2   Diagenetic facies 3   Diagenetic facies 4

Fig. 2.7 Diagenetic facies in a ramp environment with patchy ooid shoals. Pervasive dolomitization causes low porosity and high permeability (diagenetic facies 1), while limestones with moldic porosities show high porosity and low permeability (diagenetic facies 2). Vuggy porosities that have been created by meteoric dissolution (diagenetic facies 3) and hydrothermal dolomites that are related to fault and fractures caused by a salt dome are obvious. Diagenetic facies 1 and 2 follow the primary depositional settings, while 3 and 4 are independent of sedimentary environment

diagenetic-related reservoir properties are expressed with numbers, but generally they still have a qualitative nature and are not used in some parts of reservoir studies, such as static and dynamic modeling (Nader 2017). For example, the percent of each pore type can be reordered as a number in petrographical studies, but usually these numbers are not used in subsequent steps of formation evaluation. At best, the dominant pore type is taken into account.

At the microscopic scales, petrographical studies are the main source of information. Standard sheets are used for quantifying diagenetic processes (Tavakoli 2018). The amount of minerals, usually in percent, can be determined using comparison charts, point counting method or quantitative X-ray diffraction (XRD). Newly developed micro-CT scanning method is also used for determining mineral content (Nader 2017). Each pore type can be estimated using comparison charts or point counting method. Microfractures are also studied, but converting the qualitative observations to quantitative data is difficult for fractures (see Sect. 4.2 for more information). The frequency of fractures can be recorded between two depths. The 0–4 scale is

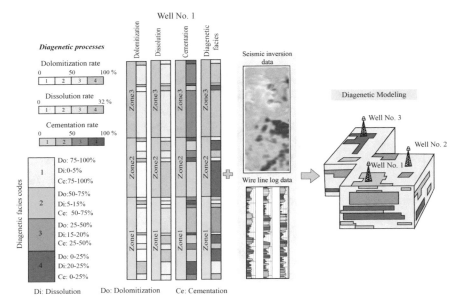

Fig. 2.8 Different stages for defining a diagenetic facies from various diagenetic processes. The units can be modeled integrating with wire-line and seismic data

usually used equivalent to the terms none, rare, common, abundant and very abundant. Cementation is a routine diagenetic process in carbonates. Isopachous, blocky and bladed cements are usually present. They are also recorded between 0 and 4. Compaction-related features such as stylolites and solution seams show the effect of overburden pressure. They can also be the result of tectonic forces. The orientation of these structures cannot be distinguished from thin sections, but their frequency between none and very abundant is recorded on standard sheets. Dolomites are routinely divided into planar and non-planar based on their shape and micro and sucrosic based on their size. Their frequency is recorded in the same range.

After data gathering, a comprehensive analysis of data is necessary before defining any diagenetic or sedimentary–diagenetic facies. For example, the relationships between diagenetic processes, their impacts and primary facies with final reservoir quality should be considered. Using a large database, Tavakoli and Jamalian (2019) considered such relationship between primary facies, dolomitization, pore types distribution and anhydrite cements. Results showed that the limy grainstone facies with moldic porosity as well as dolomitized samples with various facies have highest amount of empty spaces (Fig. 2.9). Dolomitization also has increased the porosity of the samples. In dolomitized intervals, grain- and mud-dominated facies are more frequent in high and low porosity samples, respectively (Fig. 2.10). So, the reservoir parts are composed of high porosity grainstones as well as dolomitized samples with different facies. Anhydrite cements have filled the spaces and reduce overall porosity of the samples.

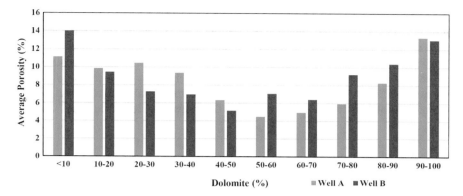

Fig. 2.9 Average porosity of the samples with various dolomite content from two wells (A and B) drilled in Permian–Triassic formations (Dalan and Kangan) of the central Persian Gulf (Tavakoli and Jamalian 2019). Samples with low and high dolomite contents have the highest porosity which shows the effect of both primary facies and secondary dolomitization on porosity evolution of this reservoir

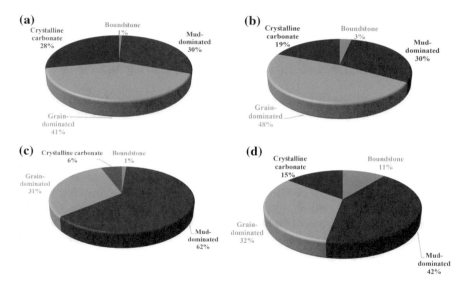

Fig. 2.10 Primary textures of the dolomitic samples from two Permian–Triassic wells in the central Persian Gulf. Samples with more than 10% porosity (**a** and **b** from wells A and B, respectively) and less than 10% porosity (**c** and **d** from wells A and B, respectively) are illustrated (Tavakoli and Jamalian 2019)

At the larger scales, diagenetic processes are quantified with other methods. The mineralogical content (e.g., volume of dolomite) can be calculated using geological maps coupled with topographic data or digital elevation models. It is noteworthy that the shape of the body should also be clearly understood to calculate the underground volume. Quantitative diagenetic data can be correlated to seismic sections (e.g., Sagan and Hart 2006). Such correlation is also possible through comparison to other petrographical data or electrofacies derived from wire-line logs (see Sect. 2.9).

2.3   CT Scan Data

Image analysis techniques are used in various parts of a reservoir evaluation. These methods try to find the relationships between petrophysical parameters and image characteristics. Various types of images including core CT scans (computed tomography scan) can be used to understand the reservoir heterogeneity. CT scan is one of the advanced medical imaging techniques with non-medical applications including core CT scanning for the petroleum industry. Using this technique, the texture and geometry of the cross-sectional images of the rocks can be examined. In this method, a thin beam of X-rays (such as a laser beam) is passed through the object from various angles. The detectors measure the X-ray intensity attenuation and convert the rays into images. Results are displayed as multiple images showing the cross-sections of the object.

Routinely, images are prepared perpendicular to the longitudinal axis of the cores. Distance between two consecutive images depends on the objectives of the project and tool's resolution. The grayscale values of a CT scan image are prepared according to the density of rock materials. Porosities with low density are recognized by black color, while matrix and allochems are white to gray. Some features can be deduced from the images by visual observation. Porosities (Fig. 2.11a–f), sedimentary structures such as cross-bedding (Fig. 2.11e, f), massive structures (Fig. 2.11g), facies changes (Fig. 2.11h) or stylolites (Fig. 2.11i) are some examples. Different methods and programs are used for image analysis in earth sciences. ImageTool (IT), ImageJ and JMicroVision are some of these programs which can extract and analyze the image features and geometry of the pores.

Heterogeneity of images can be evaluated by various methods such as point counting, image segmentation and color profiling. Some of these methods extract or calculate the image features. These features are used for understanding the heterogeneity of the sample. For example, percentage of each pore type can be calculated using point counting method. The others classify the pixels based on their gray intensity. Then, images are classified based on the dominant class of each photograph.

In addition to these methods, machine vision techniques are used for image processing and estimating the features. In recent years, intelligent systems such as neural networks, in particular deep learning, are frequently used for this purpose. These methods are also used for classification of core CT scan images (e.g., Sun et al. 2017). For example, images can be grouped into two major categories including porosity (porous samples) (Fig. 2.11a–f) and matrix (Fig. 2.11g–i).

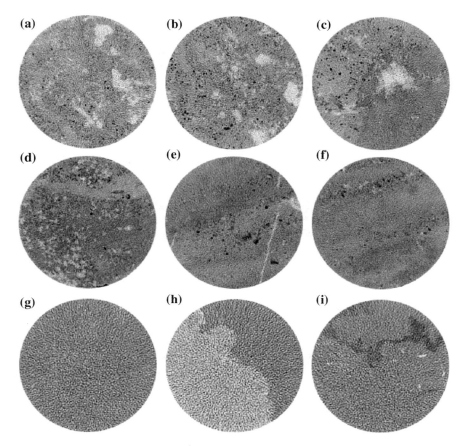

Fig. 2.11 Core CT scan images from Triassic Kangan Formation in south Iran. See text for more explanation

2.4 Porosity–Permeability Relationships

Understanding porosity–permeability relationships is one of the most important subjects in petroleum industry. In fact, a good correlation between porosity and permeability indicates a more homogeneous unit and enables the researcher to predict one parameter based on the other. These units can be correlated through the 3D space of the reservoir and enable the user to predict the oil in place (OIP) and production rate of the reservoir. The relationship between porosity and permeability of the carbonates is still a significant challenge for petroleum industry. In contrast to sandstone reservoirs which usually have a simple pore system, the pore types and their connection are very complicated in carbonates. They are also susceptible to diagenetic processes which cause major changes after precipitation. Samples with non-touching pores have low permeability. For example, moldic pores strongly increase the porosity of the sample, while it has no strong effect on permeability. Conversely, both porosity and permeability increase through dolomitization process, but its effect on permeability is more significant. Many researchers tried to introduce a pore system classification based on such relationships (e.g., Lucia 1995). From this point of view, porosities are classified into connected and disconnected. Interparticle and intercrystalline pores are the most frequent connected pores, while moldic and vuggy are dominant disconnected pore types in carbonates. Non-effective pores can be connected by some later diagenetic processes such as dolomitization or fracturing.

The degree of correlation between porosity and permeability is always a useful criterion in recognizing the degree of homogeneity in a rock type. High coefficient of determination (R^2) value shows a good correlation and so effectiveness of the used method and its ability to differentiate the samples based on the same reservoir properties. There is no known equation to relate porosity and permeability in carbonate systems, but efforts have been made to construct such relationship (e.g., Ehrenberg and Nadeau 2005). So, the R^2 value of such relationship can be calculated based on any equation. The higher value indicates the more appropriate equation.

2.4.1 Limestones

The porosity–permeability relationships in limestones depend on many parameters. The original facies and fabric as well as diagenetic processes are responsible for the goodness of fit between porosity and permeability. Originally, the grainstone facies have higher porosity and permeability because of their interparticle pore spaces. This pore system is also homogeneous. In fact, the interparticle porosity between allochems is very similar to sandstone pore systems. After deposition, they are subject to marine diagenesis. Isopachous rim cements build a rigid framework. They bind the allochems to each other. These cements are just around the grains and so they have no strong effect on porosity or permeability of the samples. Micritization is a routine process in calm environments such as lagoon depositional setting. In this process, the microspar and sparite are converted to micrite by bacterial activities. Micritization increases the amount of micrite and so the microporosity of the samples. These samples have high porosity and low permeability. The Albian–Aptian Dariyan Formation (equivalent to Shuaiba Formation in Arabian territories) in the Persian Gulf area is a good example. The samples of this formation contain no visible porosity in petrographic studies (Naderi-Khujin et al. 2016), but wire-line log data and core analysis indicate the average of 15% porosity in the samples. The micropores have been developed in meteoric diagenesis through partial dissolution of micrite particles (Tavakoli and Jamalian 2018). As microporosity is dominant pore system, a good correlation can be seen between the porosity and permeability of the samples which increases the homogeneity of the rocks. Mud-dominated samples (mudstone and wackestone) have lower primary porosity and permeability. The micrite particles fill the empty spaces. These samples can have high amounts of microporosity, which discussed previously. Recently, a review of the genesis and distribution of limestone microporosity has been presented by Hashim and Kaczmarek (2019).

After marine diagenesis, samples are usually affected by meteoric waters. These waters dissolve the unstable allochems such as aragonitic ooids and have a major impact on porosity development. Such diagenesis creates isolated molds which are not connected to each other. So, porosity increases considerably while permeability is almost constant. There is no correlation between porosity and permeability of the samples. If moldic porosities are unique pore system in the sample, this process has negligible effect on reservoir heterogeneity in microscale. A combination of various pore types causes different types of heterogeneity and also porosity–permeability relationships. Interparticle, intercrystalline or fracture porosities connect the moldic pores and so increase permeability along with porosity of the carbonate rocks. These pore types increase the geological heterogeneity of the samples and cause better porosity–permeability correlation of the samples. In contrast, separate molds or vugs may be accompanied by other disconnected pores such as micropores. Porosity significantly increases in these cases, but permeability is nearly constant. Cementations may occlude some of the larger pores and decrease the heterogeneity of the samples. This appears to have happened in Dariyan and Shuaiba formations. Large moldic pores have been developed by dissolution in early diagenetic phase. Micrites also affected by meteoric waters, and both porosity and heterogeneity increased significantly. Large pores filled with calcite cements in late diagenesis, but micropores remained empty (Ehrenberg and Walderhaug 2015). This process not only decreased the amount of pore spaces but also the heterogeneity of the formation. The high R^2 value between porosity and permeability in this formation clearly demonstrates this homogeneity.

From the above discussion, it can be concluded that the poro–perm relationships in carbonates are affected by many parameters in various ways and so their correlation is not generally good. There is no known relationship between these two parameters in carbonate reservoirs, as mentioned before. Figure 2.12 illustrates porosity–permeability data from limestone samples (more than 90% lime) from 5 Iranian reservoirs including Dalan (Late Permian), Kangan (Early–Middle Triassic), Dariyan (Albian–Aptian), Sarvak (Albian–Turonian) and Ilam (Santonian–Campanian). The best correlation can be seen in Dariyan Formation with dominant microporosity. Ilam and Sarvak formations contain microporosity, interparticle and moldic pores. Kangan and Dalan formations have the weakest correlation. Various pore types including microporosity, moldic, inter- and interaparticle, fenestral and intercrystalline pores have been identified in these formations (Abdolmaleki et al. 2016).

2.4.2 Dolomites

Dolomites and dolomitization process have different effects on porosity–permeability relationships and reservoir heterogeneity. Many of these aspects discussed previously (Sect. 2.2.2). Many researchers believe that dolomitization increases both

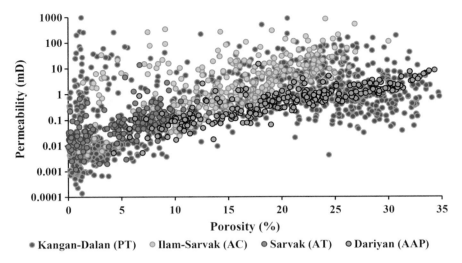

Fig. 2.12 Porosity–permeability data from various Iranian limestone formations. P: Permian, T: Triassic, A: Albian, C: Campanian, T: Turonian, AP: Aptian. See text for more explanation

porosity and permeability of carbonates. Pervasive dolomitization creates intercrystalline pores which are connected to each other. So, permeability also increases in the samples (e.g., Adam et al. 2018). A comprehensive study on the porosity–permeability distribution in carbonate reservoirs has been presented by Ehrenberg and Nadeau (2005). Considering five carbonate platforms of various ages, geographic positions and burial depths, they concluded that intercrystalline-related permeability reduces during burial. So, dolostones have higher permeability for a given porosity in shallow depth but no clear trend is visible for deeply buried reservoirs. They also found no distinct porosity–permeability trends for limestones or dolostones (Fig. 2.13).

Comparison of the four lithology types including limestone (more than 90% calcite or aragonite), dolomitic limestone (more than 50% calcite or aragonite and 10% dolomite), dolostone (more than 90% dolomite) and limy dolomite (more than 50% dolomite and 10% calcite or aragonite) from three wells of Iranian Permian–Triassic strata with about 2500–3500 m of burial depth (Fig. 2.14) showed that permeability increases along with increasing porosity even in deep buried reservoirs. It can be seen in this figure that limestone samples (Fig. 2.14a) have both low and high permeabilities for high porosities (porosity more than 20%). The number of samples with low permeability and high porosity decreases in dolomitic limestones (Fig. 2.14b). The lowest permeability is about 10 mD for the samples with higher than 20% porosity in dolostone group (Fig. 2.14c). This ratio decreases again in limy dolomites (Fig. 2.14d). This demonstrates the role of dolomitization in increasing permeability along with porosity in dolomitized reservoirs.

Heterogeneity decreases along with increasing porosity and permeability of the dolomitized samples (Fig. 2.14c).

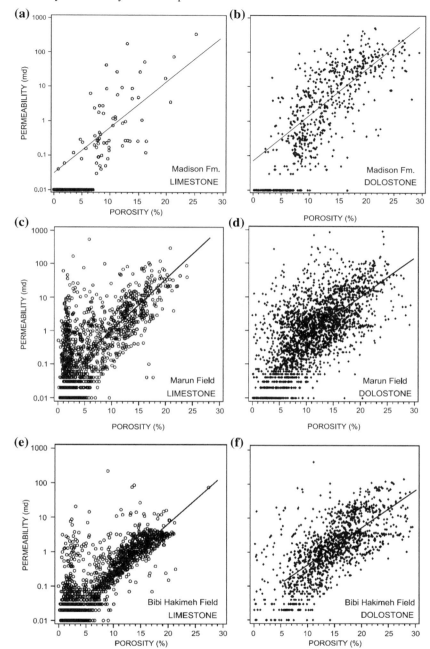

Fig. 2.13 Porosity–permeability relationships in three limestone and dolostone reservoirs. Samples are from Madison Formation outcrop sections and cores from the Big Horn and Wind River basins of Wyoming and Montana (**a, b**), Marun (**c, d**) and Bibi Hakimeh (**e, f**) fields from Iran. There are no clear relationships between porosity and permeability of the limestones or dolostones (Ehrenberg and Nadeau 2005)

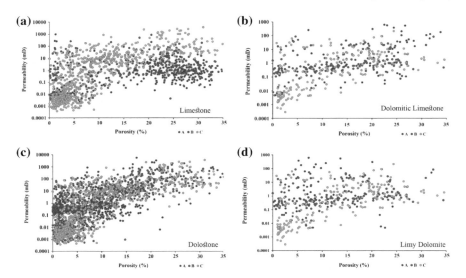

Fig. 2.14 Comparison of porosity–permeability data in four types of lithology from three Iranian wells in the central Persian Gulf. See text for more explanation

2.5   Pore Throat Sizes

Pore throat size distribution (PTSD) determines the permeability and fluid flow in a rock sample. Pore throats sizes are routinely obtained by mercury injection capillary pressure (MICP) curves. Mercury as a non-wetting phase is injected into the sample which was previously saturated by a wetting phase (usually water or air). The pressure is raised, and the sample gradually becomes saturated with respect to mercury. Pore throat sizes (PTSs) are calculated from the resultant plot of pressure versus Hg saturation. They may be plotted as a normal frequency or cumulative curves. PTSD can also be obtained by other methods such as nuclear magnetic resonance (NMR) data (e.g., Hosseini et al. 2018) or image analysis (e.g., Ferm et al. 1993).

Basic studies (Kolodizie 1980; Pittman 1992) indicate that there is a strong relationship between porosity, permeability and PTS corresponding to 35% of mercury saturation (Eq. 2.1).

$$\log r35 = 0.732 + 0.588 \log k - 0.864 \log \Phi \qquad (2.1)$$

where $r35$ is pore throat radius at 35% mercury saturation in microns, k is permeability in millidarcies, and Φ is interparticle porosity as a fraction. The coefficients have been calculated for a specific reservoir and should be recalculated for other cases. These coefficients can be determined using MICP tests and porosity and permeability measurements through a multiple regression analysis.

Rocks may be classified based on their PTSs. Classes are defined based on abrupt changes in frequency distribution of the PTSs. PTSs of 0.2, 0.5, 1, 2, 5, 15 and 60 microns have been introduced by Winland which are frequently used. It is believed that the best correlation between porosity, permeability and pore throat size is established in 35% mercury saturation. Other percentages were also introduced (e.g., Pittman 1992; Rezaee et al. 2006). This classification has been used in various fields and produced satisfactory results (e.g., Cranganu et al. 2009; Riazi 2018; Nazemi et al. 2018, 2019). Rocks within any of these groups have nearly the same PTSs and so are more homogeneous. Samples are also classified based on their r35 into megaport, macroport, mesoport and microport (Martin et al. 1997). R35 value above 10 μm represents megaport, and $r35$ of 2–10 μm corresponds to macroports. The value is between 2 and 0.5 μm in mesoport and less than 0.5 μm for microport. The samples with the same pore throats are usually scattered in studied interval, and so other methods are used for upscaling (see Sects. 3.2 and 3.4 for more information).

In limestones with moldic pores, samples are characterized by small pore throat sizes and a large pore-body size, due to the development of moldic porosity, especially grainstones with unstable aragonitic allochems. The mechanisms that may control the degree of connectivity are different. Some primary porosities are important as they create larger pore throats; microfractures and open stylolites which might also act as local fluid pathways. The pore throat distribution curves of secondary pore systems have lower sorting (Fig. 2.15a), but the efficiency is still controlled by the primary system. The scattering in pore types and pore throat network can be seen in both petrographical and SEM backscatter (Fig. 2.15b) or secondary images (Fig. 2.15c, d). Pore throat heterogeneity is still high in the sample in spite of high porosity and permeability. The dolomite samples with secondary intercrystalline porosity system (Fig. 2.16) have moderate porosity and high permeability. This is due to the fact that these secondary pores, even if not very large, are connected through a continuous pore system. So, samples are relatively homogeneous from PTS point of view (Fig. 2.16a). At larger scales in carbonate–evaporite systems, anhydrite cementation plays a significant role in determining both heterogeneity and reservoir quality (Fig. 2.16b). Increase in dolomitization rate creates microvugs which are connected to each other by intercrystalline pores (Fig. 2.16b). With increasing dolomitization rate, sample becomes more and more homogeneous (Fig. 2.17).

2.6 Sedimentary Structures

A sedimentary structure is a specific spatial relationship between size, fabric and texture of grains in a sedimentary rock. They are routinely visible with naked eye but can be seen at microscopic to macroscopic scales and indicate the physical, chemical and biological conditions of the primary depositional settings. For example, cross-bedding in an ooid shoal is sometimes visible in thin section studies. It is also visible in core scale. Sedimentary structures change the heterogeneity as well as isotropy of the reservoirs. The most common types in carbonates include stratification, cross-bedding, lenticular bodies, root traces, bioturbations, oscillation and current ripples, hummocky and swaley cross-stratification, convolute bedding and

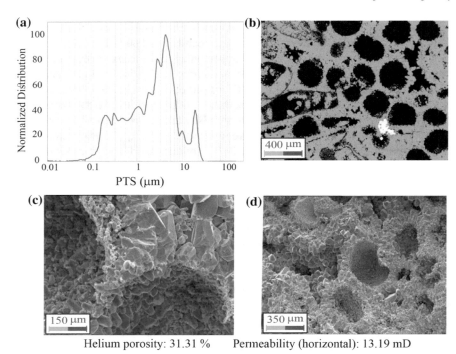

Helium porosity: 31.31 %   Permeability (horizontal): 13.19 mD

Fig. 2.15 PTSD within a limestone with high moldic and some residual primary interparticle porosities (**a**). Pore throat distribution is highly scattered indicating multiple mechanisms of pore connections. Very high porosities might locally create direct contacts between molds. Pore types and their connection in both backscattered (**b**) and secondary (**c**, **d**) SEM images can be seen

flame structures. Their role in changing the heterogeneity depends on their scale and type. For example, cross-bedded facies may have different reservoir potential from horizontally bedded facies assemblages (e.g., Anastas et al. 1998). Flow simulations on the effect of horizontal laminations and cross-beddings (Elfeki et al. 2002; Dawe et al. 2011) indicated that the microscale heterogeneities inside the macroscale units have considerable impact on fluid distribution in a porous media. The structures at various scales can be integrated to form a multiscale structure for flow simulations. A good example is cross-bedding in a thick layer. Elfeki et al. (2002) concluded that fine-scale heterogeneity causes slower and more dispersive fluid flow within a larger-scale unit. Such dispersion and slowness is because of the tortuous pathways that are generated by the fine-scale heterogeneity.

Ground penetrating radar (GPR) can be used for identification and correlation of sedimentary structures of shallow environments at mesoscale (e.g., Asprion et al. 2009; Lahijani et al. 2009; Naderi Beni et al. 2013). These structures are correlated using sedimentary environments in reservoir studies. Cross-beddings and ripples are developed in ooid shoal settings. Pseudomorphs of evaporites are found in supra-

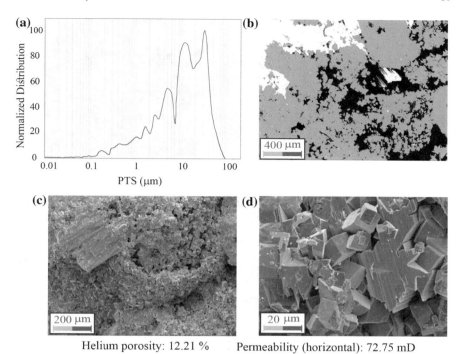

Helium porosity: 12.21 %   Permeability (horizontal): 72.75 mD

Fig. 2.16 Dolomitic sample with anhydrite patches. The pore network is characterized by the presence of primary and secondary pores, locally filled by anhydrite. So, a high scattering is visible in PTSD curve (**a**). Pores with different sizes are present (**b, c**). Microporosity was developed mainly between the dolomite crystals (**d**). The permeability is linked to the continuous pathways present at microscopic scale

and peritidal environments and can be correlated within this environment. Stroma-tolites which are precipitated in intertidal to supratidal environments contain planar or undulating millimeter-thick laminae (dark/light alternations). Bioturbations are found in intertidal, lagoon and basin sediments. Erosional surfaces indicate abrupt changes in sedimentary environments which are associated with different energy levels. Like other microscopic heterogeneities, sedimentary structures should also upscale prior to final evaluations.

2.7 Pore System Classifications

The reason for most of the classifications in a reservoir is understanding of the hetero-geneity. These classes (rock types, depositional facies, diagenetic facies and so on) are used for sample selection, conceptual and numerical modeling, flow simulations and interwell distribution of reservoir properties. Pore systems have strong effect on

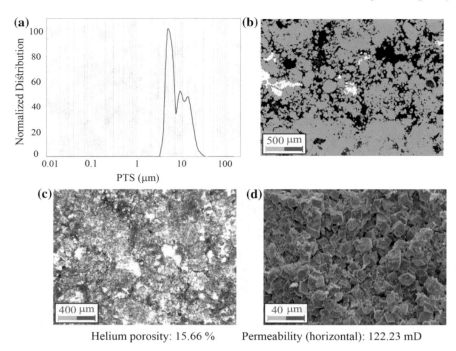

Helium porosity: 15.66 %   Permeability (horizontal): 122.23 mD

Fig. 2.17 Dolomite with anhydrite patches with low PTS scattering (**a**). The pore system is made up by intercrystalline pores and microvugs (**b**, **c**). The median pore size is low, but the permeability is high in the less cemented areas. Pores are connected via intercrystalline pores (**d**)

final reservoir properties. In fact, they control the fluid behavior (both storage and flow) and so rocks with the same pore systems exhibit the same behavior in a carbonate reservoir. In contrast to sandstones, carbonates usually have complicated and diverse pore systems. The proper choice of classification is one of the most important steps in heterogeneity evaluation of a reservoirs. An ideal pore system classification should also correlate to the facies classification system. This is performed in order to predict the pore types based on facies distribution in conceptual or numerical reservoir models. Facies models are representative of the reservoir geometry, and their distribution indicates the spatial dimension of the reservoir. So, pore types can also be predicted if they assign to a specific facies.

The Archie classification (Archie 1952) of pore types is one of the preliminary works in this field. Archie believed that a pore system classification should honor the absolute and relative permeability, capillary pressure and resistivity of the sample, rather than pore genesis and origin. He also tried to generalize his classification in order to be applicable for wellsite and field petroleum geologists. Archie explained that using these general terms, the difference between geologists' ideas about a sample will be minimum and so collected data by different experts may be compared.

The classification also includes another category for the textures of the matrix which is not the subject of this book. His classification is as follows:

Class A: The dominant pore sizes are less than 0.01 mm. They are not visible under the microscope.
Class B: Pore spaces are visible with naked eye. They are larger than 0.01 but smaller than 0.1 mm.
Class C: Visible porosities with larger than 0.1 mm and smaller than size of cuttings which is about 2 mm.
Class D: Porosities are larger than cutting size. They include vugs and large molds.

He also introduced excellent, good, fair and poor classes for 20, 15, 10 and 5% porosity in the samples, respectively. There are also some samples which are assigned to two classes (such as B/C when some large porosities are also visible, other than the fine pores). The Archie classification of pore system is still used, but with increasing the core samples, a large volume of data from thin section studies and newly developed classifications, it is not used in many cases. The advantages of Archie classification are that it can be easily applied without the need to any equipment and fast determination of the classes. Usually, size of the pores shows their formation mechanism and so classifying the samples based on their pore sizes indicates their origin. The large pores in classes C and D are secondary in origin, while class B has generally primary interparticle porosities. Intercrystalline pores are not visible in many cases and so they fall into class A. Disadvantages include generalization and lack of any indication of the porosity and permeability relationships. Micro- and intercrystalline porosities are not visible with naked eye and so they are not included in this classification. Details of the pore systems, their connection and origin are not clearly visible on hand specimens.

The most well-known pore type classification was presented by Choquette and Pray (1970). Their terms for various types of pores (Fig. 2.18) are applied in both academic studies and industrial projects. The main idea of this classification is the relationship between rock fabric and pore types. Pores are classified into three groups including fabric-selective, not fabric-selective and fabric-selective or not. Here, the fabric selectivity is relationship between the pores and solid depositional and diagenetic constituents of the rocks. For example, ooids are formed in primary depositional settings. Interparticle porosities are developed between these grains. This is a fabric-selective porosity. Calcite or anhydrite cements fill the interparticle spaces, and then ooids are dissolved in meteoric diagenetic environment. The result of this process is formation of moldic porosities. These are also fabric-selective pores because their shape and size follow the morphology of the primary constituents. In contrast, fracturing crosscuts all primary and diagenetic constituents and so it is not fabric-selective. The fabric-selective or not group may or may not follow the primary fabric of the sample. For example, burrows and borings may pass just through the rock matrix and not the grains. Both rock matrix and grains may be affected by boring organisms. So they may belong to each groups. Such porosities are called fabric-selective or not.

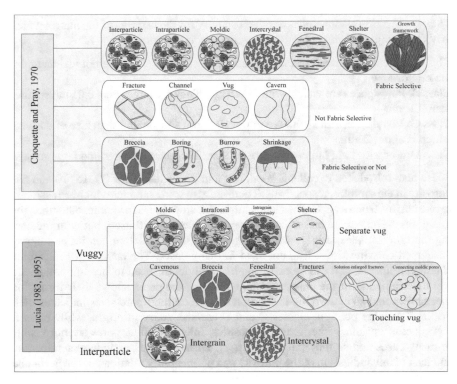

Fig. 2.18 Pore type classifications introduced by Choquette and Pray (1970) (**a**) and Lucia (1983, 1995) (**b**)

Although the terms introduced by Choquette and Pray are still used for naming the porosity types in carbonate reservoirs, relating these terms to porosity and permeability relationships and so reservoir heterogeneity is difficult. Also, it is not a genetic classification which is the basic principle for reservoir zonation through the use of sequence stratigraphic concepts. So, determined groups are less useful in determining the reservoir heterogeneity of carbonates. Lucia (1983, 1995) introduced his new classification based on the pores' connection. He defined two types of pores including interparticle (as the total intergrain and intercrystal pores) and vuggy porosities. The vugs are further subdivided into two groups including separate and touching (Fig. 2.18). The main focus of the Lucia's classification is petrophysical behavior of the rocks, and so it is useful for determining the heterogeneity of the carbonate reservoirs. The main problem is recognizing these pore types on core samples. Such differentiation needs petrographical studies of thin sections. This process is time-consuming, and thin sections are not available in many cases. However, Lucia's classification has been used in many studies and the results show that the defined classes are useful in some cases (e.g., Ehrenberg 2019; Nazemi et al. 2019; Watanabe et al. 2019).

LØnØy (2006) documented one of the best porosity classifications in terms of porosity–permeability relationships. He defined six pore types including interparticle, intercrystalline, intraparticle, moldic, vuggy and mudstone microporosity. Each pore type contains three sizes: micropores (10–50 μm), mesopores (50–100 μm) and macropores (>100 μm). Each pore size may be distributed uniformly or patchy. The R^2 value between porosities and permeabilities obtained by LØnØy for each pore type demonstrated the potential of his method in understanding reservoir heterogeneity and overcoming difficulties associate with its evaluation. The LØnØy's methodology also has the same problem. Determining pore types, pore sizes and pore distribution needs thin section availability and study of the samples under the microscope. So, this classification was limited to a few studies (e.g., Mousavi et al. 2013).

Ahr and his co-workers (Ahr and Hammel 1999; Ahr et al. 2005) introduced a genetic classification for pore types in carbonates (Fig. 2.19). In this regard, there are three pore systems including depositional, diagenetic and fracture. Hybrid types are also present which include two or three pore types. He defined a ternary plot with these three apexes. The Ahr classification is useful to differentiate the origin of porosities in carbonates and relates the pore types to the rock textures. So, it is valuable for understanding the spatial distribution of pore types but it is not designed to aid in relating porosity and permeability of the rocks. So, the fluid behavior is different for each group. For example, both inter and intraparticle porosities are the result of primary depositional fabric but fluid behavior is completely different for these two types. Interparticle porosities are commonly connected to each other and so permeability increases along with increasing porosity in these samples but intraparticle porosities are isolated and so the sample has high porosity and low permeability. This is also the case about the dissolution (moldic) and recrystallization

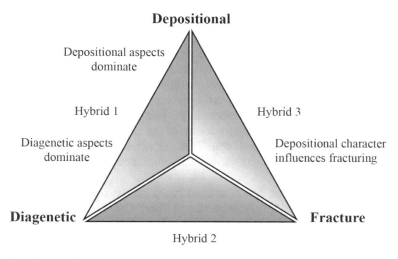

Fig. 2.19 Ahr classification for pore types in carbonate rocks

(intercrystalline) porosities. Samples with moldic pores have high porosity and low permeability, while intercrystalline pores enhance permeability rather than porosity.

Kopaska-Merkel and Mann (1991) also used ternary diagrams for pore typing in Smackover Formation in Alabama. The three apexes of their ternary diagram include moldic, interparticle and intercrystalline. Moldic porosity also contains secondary intraparticle pores that were created within the partially dissolved particles, pores between dolomite crystals within the grains and molds that have been occluded by late cementation. These pores are not connected to each other, and so Kopaska-Merkel and Mann put all of them in the same group. The intervening material could be calcite and anhydrite cements as well as micrite particles. Both interparticle and intercrystalline pores form well-connected networks, but intercrystalline pores are commonly smaller.

Kopaska-Merkel and Mann's classification integrates the genetic relationship between pores with the porosity–permeability relationships. Interparticle and intercrystalline pores form a connected network across the sample, while moldic pores are not interconnected in many cases. They also have different pore volume and pore sizes. Interparticle pores are routinely larger than intercrystalline pores. The main problem of this classification is limited pore types. They developed the ternary diagram just for Smackover Formation which has limited types of porosities.

Using ternary diagrams, Tavakoli and his co-workers (Tavakoli et al. 2011) introduced a new pore system classification (Fig. 2.20) which successfully used for classifying Permian–Triassic strata of South Pars Gas Field in Persian Gulf, the largest non-associated gas field of the world. Three apexes include depositional, fabric-selective and not fabric-selective pores. The last two classes are diagenetic in origin. Depositional pores include interparticle, fenestral and intraparticle. Fabric-selective pores include moldic and intercrystalline resulted from fabric retentive dolomitization. These pore types are not connected to each other. In fact, this is equivalent to the intraparticle pores of the Kopaska-Merkel and Mann's classification. Primary intraparticle and fenestral pores are not common in carbonate reservoirs and so have no strong effect on carbonate heterogeneity. The third apex contains the not fabric-selective pores. They include intercrystalline pores, fractures and vuggy porosities. These intercrystalline pores have been resulted from fabric-destructive dolomitization. This type of dolomitization creates well-connected pores and increases the porosity of the sample. It also enhances the porosity–permeability relationship of the rocks. This classification can be assigned to the primary rock fabric. So, it may also be used for understanding the heterogeneity of the samples and predicting spatial distribution of pore types in a reservoir. Tavakoli's classification has been successfully applied for the pore typing of various carbonate formations (e.g., Bahrami et al. 2017; Nazemi et al. 2018).

Any reservoir parameter that is defined based on the core analysis should be correlated with wire-line logs. Anselmetti and Eberli (1999) tried to determine the pore types using velocity deviation log (VDL). The VDL is calculated based on the difference between measured sonic velocity which is derived from the sonic log and synthetic velocity that is calculated from the neutron, density or their combination using Wyllie's equation (Wyllie et al. 1956) (Eq. 2.2).

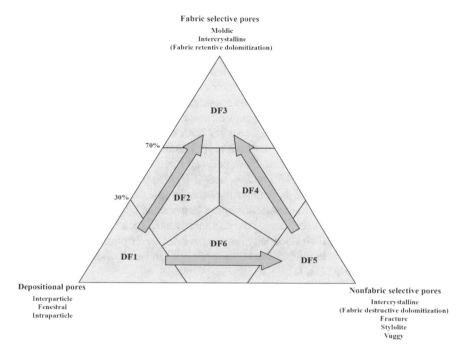

Fig. 2.20 Tavakoli pore system classification of carbonate rocks

$$VDL = \text{sonic log velocity} - \text{velocity calculated from neutron porosity or density logs} \tag{2.2}$$

There can be three possible deviations including zero, positive and negative deviations. In the case of zero deviation (±500 m/s), the real and synthetic velocities are nearly the same. So, there is no considerable porosity in the sample or the pores are connected to each other (interparticle or intercrystalline). The Wyllie's equation states that these porosities are part of the rock framework and so the sonic velocity passes through both matrix and pore spaces. These rocks have high permeability and well-connected pore network. Samples with high positive deviation contain isolated pores that are embedded in a rigid rock framework, such as moldic or intraparticle pores. Positive deviations indicate high sonic log velocity compared with neutron or density velocities. This means that the sonic wave bypasses the isolated pores and high velocities are obtained, despite high porosity. Such deviation indicates intense dissolution and precipitation of the dissolved materials as pore-filling cements. Permeability is low in such zones. The VDL has been tested in various carbonate formations, and satisfactory results were achieved in some cases (e.g., Tavakoli et al. 2011; Nazemi et al. 2018).

2.8   Rock Typing

Rock types (RT) are reservoir samples with the same petrophysical behavior. The properties of interest in a RT are fluid-related characteristics (Bear 1972; Tiab and Donaldson 2015; Tavakoli 2018). Like other classifying methods in the reservoir, RTs are also determined to overcome the heterogeneity challenges of the reservoir. In fact, all other methods such as diagenetic facies, rock fabric classes and electrofacies (see Sect. 2.9) try to define a proper RT with various points of view. Various methods have been developed to define a representative RT in a reservoir. These methods have been used, explained, classified and interpreted previously (Tiab and Donaldson 2015; Tavakoli 2018). The flow unit (FU) also has the same concept with larger scale. In fact, while RT is defined for each sample, a flow unit is defined to integrate these microscale characteristics into a larger scale which can be correlated through the reservoir. Table 2.1 shows the commonly used rock typing and flow unit determination methods, their necessary data and a brief description of each process. RTs, FUs and electrofacies use various data types but can be related to each other in a meaningful way (Fig. 2.21).

Table 2.1 Various rock typing and flow unit definition methods in carbonate reservoirs

Name	Used data	Methodology
Geological rock typing (GRT)	Facies and diagenesis data from thin sections and macroscopic core descriptions	Samples with the same geological and petrophysical properties are defined
Winland r35	Porosity, permeability and pore throat size in 35% of mercury saturation in MICP test	Samples are classified based on the same effective pore throats
Lucia rock fabric number	Interparticle porosity and permeability	Samples with the same petrophysical classes and similar grain sizes are grouped
Reservoir quality index (RQI) and flow zone indicator (FZI)	Porosity and permeability	Samples with the same permeability to porosity ratio are classified in the same groups
Lorenz plot	Porosity and permeability	Zones are characterized based on deviations from perfect equality line

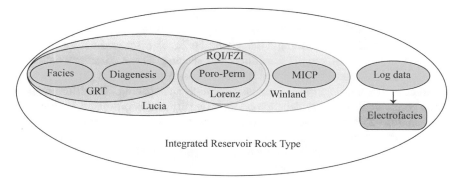

Fig. 2.21 Integrating various methods to obtain a reservoir rock type in carbonate reservoirs (Tavakoli 2018)

2.9 Electrofacies

Electrofacies are defined to provide petrophysically (routinely wire-line log) homogeneous groups (Serra 1986; Gill et al. 1993; Ye and Rabiller 2000, 2005; Lee et al. 2002; Davis 2018; Tavakoli 2018). There are various methods for determining an electrofacies, but the most commonly used is cluster analysis. The distance between each pair of data is calculated, and then, pairs with minimum distance are grouped to form a cluster. The distance between clusters is then calculated, and larger clusters are made. The process continues to integrate all data into one cluster. This process is stopped at a desired distance, and specific number of clusters are introduced.

All microscopic data should be integrated to build a unique unit to overcome the challenges of the heterogeneity in a reservoir. So, GRTs are combined with petrophysical RTs. Then, electrofacies are integrated with the results of this process and a final homogeneous unit is introduced (Fig. 2.21).

2.10 Microscopic Uncertainties

Microscopic studies of the reservoir samples start from the SEM analysis and continue with the petrographical observations of thin sections, facies and diagenetic analysis, determining the PTSD, considering the CT scan data and final reservoir rock typing. All of them are used for understanding spatial distribution of reservoir properties in field scale. These samples are just a small volume of the whole reservoir, and so, care should be taken when they are used for reservoir characterization. Petrography of thin sections needs a geological expert with enough experience. A large volume of data is collected in this process and just 1% error in distinguishing the parameters or transferring data to standard sheets resulting in wrong interpretations.

Determining the primary facies and diagenetic facies groups is a qualitative and complicated process which in part depends on the geologist's experience and idea. There are various methods for defining a diagenetic facies, as discussed before. CT scan images are taken with various resolution and spacing. These parameters strongly affect the final interpretations and even calculations. Allochems and porosities could be larger or smaller than the CT image resolution and spacing. The differentiation of the components strongly depends on their density and is presented as grayscale between 0 and 255. So, extracting the rock constituents is not easy. Pore systems have strong effect on reservoir heterogeneity and predicting other reservoir parameters in a 3D space. Various methods can be used to classify the pore types of a reservoir; each of them has its special point of view. Selection of the most suitable method based on the properties of the studied reservoir is very important. Various logs may be integrated to build a set of electrofacies. Log selection depends on the data correlation and the purpose of the study, which are different case by case. The final method for compiling all microscopic units is also a complex procedure. Combining geological and petrophysical data with a suitable method for rock typing are the most challenging part of any reservoir evaluation, especially carbonates. This challenge can be solved with the incorporation of all disciplines and use of quantitative data, as much as possible.

References

Abdolmaleki J, Tavakoli V, Asadi-Eskandar A (2016) Sedimentological and diagenetic controls on reservoir properties in the Permian-Triassic successions of Western Persian Gulf, Southern Iran. J Pet Sci Eng 141:90–113

Adam A, Swennen R, Abdulghani W, Abdlmutalib A, Hariri M, Abdulraheem A (2018) Reservoir heterogeneity and quality of Khuff carbonates in outcrops of central Saudi Arabia. Mar Pet Geol 89:721–751

Ahr WM, Hammel B (1999) Identification and mapping of flow units in carbonate reservoirs: an example from Happy Spraberry (Permian) Field, Garza County, Texas USA. Energy Explor Exploit 17:311–334

Ahr WM, Allen D, Boyd A, Bachman HN, Smithson T, Clerke EA, Gzara KBM, Hassall JK, Murty CRK, Zubari H, Ramamoorthy R (2005) Confronting the carbonate conundrum. Schlumberger Oil field Rev 1:18–29

Anastas AS, James NP, Nelson CS, Dalrymple RW (1998) Deposition and textural evolution of cool-water limestones: outcrop analog for reservoir potential in cross-bedded calcitic reservoirs. AAPG Bull 82:160–180

Anselmetti FS, Eberli GP (1999) The velocity-deviation log: a tool to predict pore type and permeability trends in carbonate drill holes from sonic and porosity or density logs. AAPG Bull 83:450–466

Archie GE (1952) Classification of carbonate reservoir rocks and petrophysical considerations. AAPG Bull 36(2):278–298

Asprion U, Westphal H, Nieman M, Pomar L (2009) Extrapolation of depositional geometries of the Menorcan Miocene carbonate ramp with ground-penetrating radar. Facies 55:37–46

Bahrami F, Moussavi-Harami R, Khanehbad M, Gharaie MM, Sadeghi R (2017) Identification of pore types and pore facies for evaluating the diagenetic performance on reservoir quality: a case study from the Asmari Formation in Ramin Oil Field, SW Iran. Geosci J 21:565–577

Bear J (1972) Dynamics of fluids in porous media. American Elsevier Publishing Co., New York

Bruna PO, Lavenu APC, Matonti C, Bertotti G (2019) Are stylolites fluid-flow efficient features? J Struct Geol 125:270–277

Carnell AJH, Wilson MEJ (2004) Dolomites in SE Asia–varied origins and implications for hydrocarbon exploration. Geol Soc Lond Spec Publ 235:255–300

Choquette PW, Pray LC (1970) Geologic nomenclature and classification of porosity in sedimentary carbonates. AAPG Bull 54(2):207–250

Cranganu C, Villa MA, Saramet M, Zakharova N (2009) Petrophysical characteristics of source and reservoir rocks in the Histria basin, western Black Sea. J Pet Geol 32:357–372

Cui Y, Wang G, Jones SJ, Zhou Z, Ran Y, Lai J, Li R, Deng L (2017) Prediction of diagenetic facies using well logs—a case study from the upper Triassic Yanchang Formation, Ordos Basin, China. Mar Pet Geol 81:50–65

Davis JC (2018) Electrofacies in reservoir characterization. In: Daya Sagar B, Cheng Q, Agterberg F (eds) Handbook of mathematical geosciences. Springer, Cham

Dawe RA, Caruana A, Grattoni CA (2011) Immiscible displacement in cross-bedded heterogeneous porous media. Transp Porous Med 87:335–353

Dunham RJ (1962) Classification of carbonate rocks according to depositional texture. In: Ham WE (ed) Classification of carbonate rocks. AAPG Memoir 1, Oklahoma

Ehrenberg SN (2019) Petrophysical heterogeneity in a lower cretaceous limestone reservoir, onshore Abu Dhabi, United Arab Emirates. AAPG Bull 103:527–546

Ehrenberg SN, Nadeau PH (2005) Sandstone vs. carbonate petroleum reservoirs: a global perspective on porosity-depth and porosity-permeability relationships. AAPG Bull 89:435–445

Ehrenberg SN, Walderhaug O (2015) Preferential calcite cementation of macropores in microporous limestones. J Sediment Res 85:780–793

Elfeki AM, Dekking FM, Bruining J, Kraaikamp C (2002) Influence of fine-scale heterogeneity patterns on large-scale behavior of miscible transport in porous media. Pet Geosci 8:159–165

Embry AF, Klovan JE (1971) A Late Devonian reef tract on Northeastern Banks Island, NWT. Can Petro Geol Bull 19:730–781

Ferm JB, Ehrlich R, Crawford GA (1993) Petrographic image analysis and petrophysics: analysis of crystalline carbonates from the Permian basin, west Texas. Carbonates Evaporites 8:90–108

Flugel E (2010) Microfacies of carbonate rocks, analysis, interpretation and application. Springer, Berlin

Friedman GM (1965) Terminology of crystallization textures and fabrics in sedimentary rocks. J Sediment Pet 35:643–655

Gill D, Shomrony A, Fligelman H (1993) Numerical zonation of log suites and logfacies recognition by multivariate clustering. AAPG Bull 77:1781–1791

Gressly A (1838) Observations géologiques sur le Jura soleurois, vol 1. Imprimerie de Petitpierre (in French)

Hashim MS, Kaczmarek SE (2019) A review of the nature and origin of limestone microporosity. Mar Pet Geol 107:527–554

Hosseini M, Tavakoli V, Nazemi M (2018) The effect of heterogeneity on NMR derived capillary pressure curves, case study of Dariyan tight carbonate reservoir in the central Persian Gulf. J Pet Sci Eng 171:1113–1122

Huang SJ, Huang KK, Lu J, Lan YF (2014) The relationship between dolomite textures and their formation temperature: a case study from the Permian-Triassic of the Sichuan Basin and the lower Paleozoic of the Tarim Basin. Pet Sci 11:39–51

Koehn D, Rood MP, Beaudoin N, Chung P, Bons PD, Gomez-Rivas E (2016) A new stylolite classification scheme to estimate compaction and local permeability variations. Sed Geol 346:60–71

Kolodizie SJ (1980) Analysis of pore throat size and use of the Waxman-Smits equation to determine OOIP in Spindle Field, Colorado. SPE paper 9382 presented at the 1980 SPE annual technical conference and exhibition, Dallas, Texas

Kopaska-Merkel DC, Mann SD (1991) Pore facies of Smackover carbonate reservoirs in southwest Alabama. Gulf Coast Ass Geol Soc Trans 41:374–382

Lahijani HAK, Rahimpour-Bonab H, Tavakoli V, Hosseindoost M (2009) Evidence for late Holocene highstands in Central Guilan-East Mazanderan, South Caspian coast, Iran. Quatern Int 197:55–71

Lai J, Fan X, Pang X, Zhang X, Xiao C, Zhao X, Han C, Wang G, Qin Z (2019) Correlating diagenetic facies with well logs (conventional and image) in sandstones: the Eocene-Oligocene Suweiyi Formation in Dina 2 Gasfield, Kuqa depression of China. J Pet Sci Eng 174:617–636

Lee SH, Kharghoria A, Datta-Gupta A (2002) Electrofacies characterization and permeability predictions in complex reservoirs. SPE Reserv Eval Eng 5:237–248

LØnØy A (2006) Making sense of carbonate pore systems. AAPG Bull 90:1381–1405

Lucia FJ (1983) Petrophysical parameters estimated from visual description of carbonate rocks: a field classification of carbonate pore space. J Pet Tech 35:626–637

Lucia FJ (1995) Rock fabric/petrophysical classification of carbonate pore space for reservoir characterization. AAPG Bull 79:1275–1300

Lucia FJ, Major RP (1994) Porosity evolution through hypersaline reflux dolomitization. In: Purser B, Tucker M, Zenger D (eds) Dolomites: International Association of Sedimentologists Special Publications 21345–360

Machel HG (2004) Concepts and models of dolomitization—a critical reappraisal. In: Braithwaite C, Rizzi G, Darke G (eds) The geometry and petrogenesis of dolomite hydrocarbon reservoirs. Geological Society, London, Special Publications 235:7–63

Martin AJ, Solomon ST, Hartmann DJ (1997) Characterization of petrophysical flow units in carbonate reservoirs. AAPG Bull 81:734–759

Mehrabi H, Mansouri M, Rahimpour-Bonab H, Tavakoli V, Hassanzadeh M (2016) Chemical compaction features as potential barriers in the Permian-Triassic reservoirs of South Pars Field, Southern Iran. J Pet Sci Eng 145:95–113

Mousavi M, Prodanovic M, Jacobi D (2013) New classification of carbonate rocks for process-based pore-scale modeling. SPE J 18:243–263

Nader FH (2017) Multi-scale quantitative diagenesis and impacts on heterogeneity of carbonate reservoir rocks. Springer, Cham, Switzerland

Naderi Beni A, Lahijani H, Moussavi Harami R, Leroy SAG, Shah-Hosseini M, Kabiri K, Tavakoli V (2013) Development of spit-lagoon complexes in response to Little Ice Age rapid sea-level changes in the central Guilan coast, South Caspian Sea, Iran. Geomorphology 187:11–26

Naderi-Khujin M, Seyrafian A, Vaziri-Moghaddam H, Tavakoli V (2016) Characterization of the Late Aptian top-Dariyan disconformity surface, offshore SW Iran: a multi-proxy approach. J Pet Geol 39:269–286

Nazemi M, Tavakoli V, Rahimpour-Bonab H, Hosseini M, Sharifi-Yazdi M (2018) The effect of carbonate reservoir heterogeneity on Archie's exponents (a and m), an example from Kangan and Dalan gas formations in the central Persian Gulf. J Nat Gas Sci Eng 59:297–308

Nazemi M, Tavakoli V, Sharifi-Yazdi M, Rahimpour-Bonab H, Hosseini M (2019) The impact of micro-to macro-scale geological attributes on Archie's exponents, an example from Permian-Triassic carbonate reservoirs of the central Persian Gulf. Mar Pet Geol 102:775–785

Pittman ED (1992) Relationship of porosity and permeability to various parameters derived from mercury injection-capillary pressure curves for sandstone. AAPG Bull 72:191–198

Purser BH, Brown A, Aissaoui DM (1994) Nature, origins and evolution of porosity in dolomite. In: Purser B, Tucker M, Zenger D (eds) Dolomites. International Association of Sedimentologists Special Publications 21:283–308

Rezaee MR, Jafari A, Kazemzadeh E (2006) Relationship between permeability, porosity and pore throat size in carbonate rocks using regression analysis and neural networks. J Geophys Eng 3:370–376

Riazi Z (2018) Application of integrated rock typing and flow units identification methods for an Iranian carbonate reservoir. J Pet Sci Eng 160:483–497

Sagan JA, Hart BS (2006) Three-dimensional seismic-based definition of fault-related porosity development: Trenton-Black River interval, Saybrook, Ohio. AAPG Bull 90:1763–1785

Saller AH (2004) Palaeozoic dolomite reservoirs in the Permian Basin, SW USA: stratigraphic distribution, porosity, permeability and production. Geol Soc Lond Spec Publ 235:309–323

Saller AH, Henderson N (1998) Distribution of porosity and permeability in platform dolomites: insight from the Permian of west Texas. AAPG Bull 82:1528–1550

Saller AH, Henderson N (2001) Distribution of porosity and permeability in platform dolomites: insight from the Permian of west Texas: reply. AAPG Bull 85:530–532

Serra O (1986) Fundamentals of well log interpretation, vol 2. The interpretation of logging data. Elsevier, Amsterdam

Sibley DF, Gregg JM (1987) Classification of dolomite rock textures. J Sed Pet 57:967–975

Sun H, Vega S, Tao G (2017) Analysis of heterogeneity and permeability anisotropy in carbonate rock samples using digital rock physics. J Pet Sci Eng 156:419–429

Tavakoli V (2018) Geological core analysis: application to reservoir characterization. Springer, Cham, Switzerland

Tavakoli V, Jamalian A (2018) Microporosity evolution in Iranian reservoirs, Dalan and Dariyan formations, the central Persian Gulf. J Nat Gas Sci Eng 52:155–165

Tavakoli V, Naderi-Khujin M, Seyedmehdi Z (2018) The end-permian regression in the western tethys: sedimentological and geochemical evidence from offshore the Persian Gulf, Iran. Geo-Mar Lett 38(2):179–192

Tavakoli V, Jamalian A (2019) Porosity evolution in dolomitized Permian-Triassic strata of the Persian Gulf, insights into the porosity origin of dolomite reservoirs. J Pet Sci Eng 181:106191

Tavakoli V, Rahimpour-Bonab H, Esrafili-Dizaji B (2011) Diagenetic controlled reservoir quality of South Pars gas field, an integrated approach. C R Geosci 343:55–71

Tiab D, Donaldson EC (2015) Petrophysics, theory and practice of measuring reservoir rock and fluid transport properties. Gulf Professional Publishing, Houston

Tucker M, Wright VP (1990) Carbonate sedimentology. Blackwell Scientific Publications, Oxford

Wang J, Cao Y, Liu K, Liu J, Kashif M (2017) Identification of sedimentary-diagenetic facies and reservoir porosity and permeability prediction: an example from the Eocene beach-bar sandstone in the Dongying Depression, China. Mar Pet Geol 82:69–84

Warren J (2000) Dolomite: occurrence, evolution and economically important associations. Earth Sci Rev 52:1–81

Watanabe N, Kusanagi H, Shimazu T, Yagi M (2019) Local non-vuggy modeling and relations among porosity, permeability and preferential flow for vuggy carbonates. Eng Geol 248:197–206

Wyllie MR, Gregory AR, Gardner GHF (1956) Elastic wave velocities in heterogeneous and porous media. Geophysics 21:41–70

Ye S, Rabiller P (2000) A new tool for electro-facies analysis: multi-resolution graph-based clustering. In: SPWLA 41st annual logging symposium, Dallas, Texas, pp 14–27

Ye S, Rabiller P (2005) Automated electrofacies ordering. Petrophysics 46:409–423

Zou C, Tao S, Zhou H, Zhang X, He D, Zhou C, Wang L, Wang X, Li F, Luo P, Yuan X (2008) Genesis, classification, and evaluation method of diagenetic facies. Pet Explor Dev 35:526–540

Chapter 3
Mesoscopic Heterogeneity

Abstract Microscopic homogeneous unites which are defined based on the fine-scale data are combined to form larger-scale units. Facies are grouped together to create a facies group and facies belts. These groups are interpolated between the wells using the concept of facies models. These models are constructed by comparing various sedimentary environments from different parts of the world. They act as a template for comparison, future observations, distribution of geological properties in space and time and also for understanding the primary physical, chemical and biological conditions of the depositional settings. While facies, facies groups and facies belts indicate the geological heterogeneity of the reservoir, hydraulic flow units divide the intervals into homogeneous groups from petrophysical point of view. These units are defined using various methods which are still evolving. Petrophysical characteristics are integrated with facies models and properties are upscaled into a coarser grid suitable for numerical reservoir modeling. The cyclicities of the facies and properties indicate the periodicity of the primary depositional conditions which, in turn, are used for heterogeneity analysis of the strata. These cyclicities are defined using sequences in most cases. Both geological (facies and diagenesis) and petrophysical properties are correlated in the framework of sequence stratigraphic units in the reservoir. Rocks between the boundaries of the sequences are genetically related and so have similar primary geological and petrophysical characteristics. With respect to these similarities, their diagenesis are also the same in most cases. So, defining facies, facies groups, facies models, hydraulic flow units and recognizable genetically related units at various scales helps to overcome the challenges of heterogeneity in carbonate reservoirs.

3.1 Sedimentary Environments

Facies are the main building blocks of the reservoirs' body from sedimentological point of view. Also, other reservoir characteristics such as diagenesis or petrophysical parameters usually follow the primary facies of the rocks. So, understanding both horizontal and vertical facies distributions plays a fundamental role in any formation

© The Author(s), under exclusive license to Springer Nature Switzerland AG 2020   53
V. Tavakoli, *Carbonate Reservoir Heterogeneity*,
SpringerBriefs in Petroleum Geoscience & Engineering,
https://doi.org/10.1007/978-3-030-34773-4_3

evaluation. The concept of sedimentary environment as a media for sediment accumulation has been used for many years by sedimentologists. The spatial relationships between facies, their sequence and resulting vertical accumulation were first introduced by (Walker 1984). He stated that the facies and sedimentary environments which are observed beside each other are superimposed during the transgression and regression cycles. This means that two consecutive facies in a vertical column belong to two adjacent sedimentary environments. Although the law should be used with care, it provides the basis for reconstructing sedimentary environments and spatial distribution of the facies based on their sedimentary environments. It is worth mentioning that erosional surfaces or any other discontinuity in the rock record disturb this relationship. A facies model (Walker 1984) is representing the relationships between facies and combining them based on the concept of sedimentary environments. Facies models can be presented by a sequence of facies, conceptual models, block diagrams and graphs. The concept of facies model, which is usually expressed as sedimentary environments, is used to predict the spatial properties of the reservoir rocks, at least from the sedimentary point of view. As mentioned, many other properties follow this distribution. The introduction of the facies models is one of the main revolutions in the science of petroleum geology.

There are four important functions for a facies model (Walker 1984) including (1) it is used for purposes of comparison. This means that any facies model must act as a template or standard. Any sequence of facies is compared with a standard facies model and so follows its geometry, reservoir properties and rock textures. A carbonate ramp is a good example. A sequence of peritidal mudstone, lagoonal wackestone and some sparse ooid grainstones of the shoal without any evidence of a large and extensive barrier clearly indicates a ramp environment. Other evidence such as sharp change in slope and sliding and slumping structures should also be present. After recognizing the environment, the ramp template can be used for distributing the properties. Other examples include low reservoir quality as well as blanket geometry in peritidal environments, mud-dominated microporous lagoon sediments and meteoric dissolution in aragonitic ooids of the shoals. (2) It acts as a framework for future observations. A distinct facies model has known primary and secondary characteristics which can be used for other environments. (3) A facies model predicts the distribution of geological properties in the reservoir. (4) It is used for interpretation of the primary physical, geochemical and biological conditions of the rocks. Regarding these characteristics, a facies model and its resulting sedimentary environment can be used for understanding and predicting the heterogeneity of a reservoir at mesoscale.

Defining a sedimentary environment enables the construction of both conceptual and numeric 3D facies models. Each sedimentary environment has its specific geometry, and so the spatial distributions of facies and the related petrophysical properties are understood. The result is a facies model which can be used for both quantitative and qualitative evaluation of reservoir properties.

An important point in defining a facies model and using it as a basis for reconstructing the sedimentary environment as well as predicting reservoir properties is that the carbonates are born, not made. They are born by living organisms, and as organisms have changed during the geological times, the nature of the carbonate facies and the corresponding sedimentary environments also have been transformed during geological periods. Epeiric ramps are good examples. These extensive shallow inland seas have covered the continents during transgressive periods of sea level. The photic zone was very extensive, and so, a large area was covered by the carbonate factory. Wide facies belts were developed in these shallow seas. The subenvironments were hundreds of kilometers in length and width. The Permian–Triassic Dalan Formation equivalent to Khuff Formation in Oman, Bahrain, Saudi Arabia and Kuwait, Chia Zairi in Iraq and Bih and Hagil in United Arab Emirates has been deposited in such environment. This formation hosts the largest gas reservoir of the world.

Defining the facies and determining a facies model or sedimentary environment needs geological core studies. Regarding the Walter's law, a conceptual model can be introduced by geological studies of at least one well core, but building a numerical 3D model needs a reasonable number of cores in different wells of a field. Cores are not available in many cases or at least are not enough to reconstruct such model. So, electrofacies concept (Sect. 2.9) is used instead. They are defined based on the wireline logs which are routinely available from many wells. Unfortunately, standard facies models are not available for these electrofacies at this time. In fact, defining a model for a petrophysical property which is controlled by many variables is not possible in many cases. So, propagating the properties is based on the mathematical formulas. These formulas ignore geological heterogeneities and so are not as reliable as real facies models in reservoir studies.

The first step for determining depositional sedimentary environment is facies characterization. Facies are grouped to form a facies group. A facies group is a combination of similar facies in terms of their depositional properties (Tavakoli 2018). Facies groups, in turn, are combined to form a facies zone or belt using the same conditions. For example, peloid wackestone, bioclast peloid wackestone and peloid bioclast wackestone can be grouped into the peloid/bioclast wackestone facies group. This facies group along with other groups such as fossiliferous mudstone and peloid/bioclast packstone belong to lagoonal facies zone. These zones are constituents of a depositional environment. In this example, peritidal and lagoonal zone in combination with shoal and open marine facies zones forms a ramp depositional setting.

There are various types of carbonate environments, but few of them are susceptible to petroleum accumulation. These are marine shallow depositional settings which are main carbonate-producing environments. They include shelves, ramps, platforms and atolls. There are two types of ramps including homoclinal and distally steepened ramps (Read 1982). Homoclinal ramps are extended from shallow nearshore setting to deep offshore basin without a major break. They have very gentle (usually less than 1 degree) depositional slope. Distally steepened ramps are characterized by a distinct break at the deep part of the ramp environment. Shelves include rimmed and non-rimmed carbonates (Ginsburg and James 1974). Rimmed shelves are flat-topped carbonates with a rim in their outer ridge. This rim can be formed by reefs, shoals or islands. Non-rimmed carbonates have no effective rim in their morphology to absorb the waves' energy. So, they are similar to ramps with an incomplete barrier. Platforms include epeiric, isolated and drowned. Epeiric platforms (Shaw 1964) were characterized by very extensive shallow environments which has no counterpart in modern depositional settings. Isolated platforms are detached shallow-water carbonates generally with steep margins into deep water. Almost all carbonate depositional settings may be drowned due to subsidence, rapid sea-level rise or change in carbonate productivity as the result of environmental stress. Oceanic atolls are formed by volcanos rising hundreds of meters from the sea floor. The crest is close to the water level, and the surrounding areas are covered by deep water.

Facies types in various depositional environments are limited and gradually change with increasing the water depth. Flugel (2010) introduced the standard microfacies for both ramps and rimmed shelves. Sedimentary environments can be reconstructed based on the sequence of these facies using Walter's law. These environments transform to each other through time. Rapid sea-level rise cause drowning of the environment and rimmed shelves are converted to ramps. In contrast, sea-level fall causes subaerial exposure which also may change the environment. High siliciclastic inputs shut down the living organisms and cause major changes in the depositional settings. In geological time, plate movements change temperature, salinity and oceanic currents. These variations have strong effects on secreting organisms. The result is major changes in carbonate production in different parts of the environment.

3.2 Hydraulic Flow Units

Defining homogeneous units with respect to fluid flow and storage in carbonate reservoirs was the focus of many studies. Bear (1972) defined a hydraulic flow unit (HFU) as a volume with the same geological and petrophysical properties. Ebanks (1987) has the same definition, but he believed that such unit must be mappable at the reservoir scale. Hear et al. (1984) stated that this unit should be laterally continuous. Gunter et al. (1997) presented a same definition. Tiab and Donaldson (2015) stated that an HFU is a geological/engineering unit that is defined to describe the storage and flow properties of the reservoir zones. Tavakoli (2018) concluded from the all above definitions that a rock type and a flow unit are different in the scale of investigation. In fact, rock types are determined at the core scale, while HFUs are defined at reservoir scale. So, HFUs can be correlated between the wells.

Amaefule et al. (1993) introduced the ratio of permeability to effective porosity as a good indicator of HFUs. They defined the reservoir quality index (RQI) (Eq. 3.1) and flow zone indicator (FZI) (Eqs. 3.2 and 3.3) which were used in many studies to evaluate the heterogeneity of the reservoirs.

$$RQI\ (\mu m) = 0.0314\ (K/\Phi)^{0.5} \tag{3.1}$$

$$FZI = RQI/\Phi_z \tag{3.2}$$

$$\Phi_z = \Phi/(1 - \Phi) \tag{3.3}$$

where Φ is fractional porosity, K is permeability in mD, and Φ_z is normalized porosity.

As mentioned, many researchers applied the Amaefule formulas in their works (e.g., Al-Dhafeeri and Nasr-El-Din 2007; Tavakoli et al. 2011; Rahimpour-Bonab et al. 2012; Arifianto et al. 2018; Nabawy et al. 2018; Riazi 2018; Liu et al. 2019; Soleymanzadeh et al. 2019). Some of these researchers believe that the method can be applied successfully in carbonate reservoirs (Fig. 3.1).

Mirzaei-Paiaman et al. (2015, 2018, 2019) divided rock typing methods into two classes. A petrophysical static rock type (PSRT) which was defined as a group of rocks with similar primary drainage capillary pressure curves. So, mercury injection capillary pressure (MICP) tests are necessary for determining these rock types. They also defined a petrophysical dynamic rock type (PDRT) (equivalent to HFU) as a group of rocks with similar fluid flow behavior. This is again a reservoir engineering definition. Knowledge of HFUs of PDRTs is needed to identify more prolific intervals. They argued that these two classes of rocks are not necessarily identical. This is not in agreement with the idea of combining all similar geological, petrophysical and engineering characteristics into one unit. It is obvious that defining such unit is not easy and many inconsistencies in nature and ranges of data complicate this process

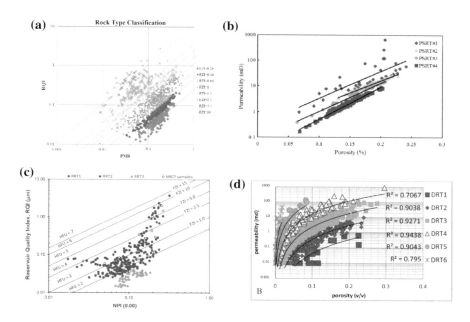

Fig. 3.1 Application of FZI method in different carbonate formations of the world. Normalized porosity versus permeability in Sarvak (Cenomanian) formation in Middle East (**a**), porosity–permeability plot of Mishrif Formation in Missan oilfield (Iraq) (**b**), Rudeis Formation (Lower Miocene) from Gulf of Suez (**c**) and porosity–permeability plot of Asmari Formation (Oligo–Miocene, Zagros) (**d**) (compiled from Riazi (2018), Liu et al (2019), Nabawy et al. (2018) and Farshi et al. (2019) from **a** to **d**, respectively)

but the ultimate goal is defining one unit which can be used in integrated static and dynamic numerical reservoir modeling.

FZI* or FZI-Star was introduced by Mirzaei-Paiaman et al. for determining the HFUs which is calculated based on the permeability–porosity ratio (Eq. 3.4) or using a multiple linear regression equation form (Eq. 3.5). From their definition, it can be concluded that FZI* and RQI of Amaefule are the same. They also believed that FZI is a function of grain size rather than pore throat size.

$$FZI^* = RQI \ (\mu m) = 0.0314 \ (K/\Phi)^{0.5} \tag{3.4}$$

$$\log(FZI^*) = -1.50307 + 0.5\log(K) - 0.5\log(\Phi) \tag{3.5}$$

Izadi and Ghalambor (2013) defined two new indices including modified reservoir quality index (MRQI) (Eq. 3.6) and modified flow zone indicator (MFZI) based on the Poiseuille and Darcy equations (Eq. 3.7) as follows:

$$MRQI = 0.0314 \ (K/\Phi)^{0.5}(1 - Sw_{ir})^{0.5} \tag{3.6}$$

$$MFZI = MRQI/\Phi_z(1 - Sw_{ir})^{0.5} \tag{3.7}$$

Although the Izadi and Ghalambor used the sandstone samples to obtain their equations, Mirzaei-Paiaman et al. (2018) used their indices for Albian–Cenomanian Ilam and Sarvak carbonate reservoirs. It is worth mentioning that these methods need further verification by using new data from other regions. Nazari et al. (2019) applied the FZI* method to tight carbonates of Upper Permian Dalan Formation in the central Persian Gulf (Fig. 3.2). They concluded that the method differentiates tight samples from other parts of the K3 reservoir unit in this formation.

3.3 Cyclicity

Repeated cycles of carbonate facies are deposited during geological time. This hierarchy is important because they are used for systematic division of strata. Understanding causes and consequences of this hierarchy is also important because an ordered forcing-mechanism is responsible for generation of these repetitions. The main control on cyclicity of sediments is attributed to changes in accommodation which, in turn, is the result of climate variations.

The repetitions and cyclic pattern of changes is deduced from similarities in sediments' properties and stratal surfaces. These bounding surfaces should be correlatable at regional scales. According to Lerat et al. (2000), the criterion for repetition is the extent of changes within a cycle. So, minor cyclic changes create smaller scale cycles, while major changes in depositional setting form the larger cycles.

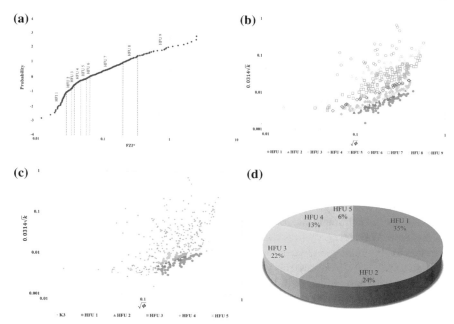

Fig. 3.2 Results of rock typing based on FZI* method in Permian Dalan Formation in the central Persian Gulf. Probability distribution of FZI* and determining the limit of each rock type (**a**), plot of 0.0314 $\sqrt{k}$ versus $\sqrt{\Phi}$ on a log–log plot to show the distribution of the samples (**b**), illustrating the tight samples (K3) on the same graph (**c**) and the percentage of tight samples in each HFU (**d**) (from Nazari et al. 2019)

Interpretation of cycles can be misleading due to the condensation, missing or major changes in depositional conditions (physical, chemical or biological). During maximum flooding, deepwater basin is starved and sediment accumulation is slow or there is no deposition. So, thin layers are developed which usually cannot be used for understanding cyclic variations in sedimentary basins. During lowstands, sediments bypass some parts of the basin or previously deposited strata are eroded and transported to the deeper parts of the basin.

Sequence stratigraphy has been used for many years to understand the cyclic patterns of sediments in wells or outcrops. Sequence stratigraphic models have been developed to describe and interpret the changes in depositional cycles. Sequence is the main building block of the sequence stratigraphic analysis. Other divisions depend on the used model, but the sea-level changes (fall and rise) are model-independent. A sequence is defined as a cycle of sediments between a successive sea-level fall and rise. Various authors placed the sequence boundary in different positions of the sea-level fluctuations. Anyway, the cyclic pattern of fall and rise is always repeated at various scales. These cycles are considered as homogeneous units at mesoscopic scales because their constituent facies have genetic relationship with each other. This means that physical, chemical and biological conditions at the time of deposition or precipitation were nearly the same for all of them. So, they have similar reservoir properties and are homogeneous at larger scales. There are different models to define a sequence and its constituent systems tracts. The appropriate model is selected based on the available data and heterogeneity of the reservoir.

Mathematical methods have been developed to indicate the cyclicity of sediments. Semivariograms (SV), Fourier analysis (FA) and more recently wavelet transformation (WT) have been used for extracting cyclicity of sediments. The SVs are routinely used for detecting spatial continuity of a property in reservoir modeling (e.g., Eltom et al. 2013), but SV of petrophysical data can also detect periodicities in reservoirs (e.g., Jennings et al. 2000; Jensen et al. 2000). The SV measures the degree of similarity between two samples regarding their separation distance. The Fourier transform is usually used for spectral analysis. It also has been utilized for detecting sedimentary cycles in reservoirs (e.g., Herbert and Fischer 1986; Schwarzacher 1993; Ellwood et al. 2008). In both methods, localization of the events in their positions (depth) is not possible. Wavelet analysis is one of the most spectacular tools in reservoir studies to detect the boundaries between the cycles and cyclicity of the events. Wavelet is an oscillation with respect to a series (which is depth in reservoir studies). Its amplitude starts from zero and returns to it. In contrast to Fourier transform which uses smooth sinusoid waves, a wavelet can be irregular. There are many standard wavelets, but any wavelet can be defined by user. The shifted and scaled versions of the selected wavelet are compared with the data series, and similarities are presented by wavelet coefficients. The maximum deviation is located at discontinuities surfaces. In fact, in reservoir studies, properties change with depth and so a continuous depth series is generated. Good examples are wire-line logs which are recorded according to depth. Then, wavelet coefficients are used for distinguishing discontinuities and cyclicities in properties. Like Fourier, wavelet analysis is a windowing technique but with varies window size. The window size increases (frequency decreases) while resolving the low-frequency oscillations of the signal. Many researchers have used this technique to detect subsurface boundaries and cyclicity in properties. Prokoph and Agterberg (2000) used gamma-ray (GR) log data and Morlet wavelet analysis to detect discontinuities and locating sedimentary cycles. Comparing the ratio of the GR log cycles with the Milankovitch fluctuations, they concluded that climatic cycles were responsible for changes in sedimentation of the Egret member, offshore Canada. They introduced a step-by-step procedure for analysis depth-series data by wavelet transform (Fig. 3.3). Wavelet analysis have also been used for other purposes such as distinguishing the pattern of Milankovitch cyclicity (e.g., Prokoph and Thurow 2000), isotope analysis (e.g., Prokoph et al. 2008) and exploring the orbital cycles (e.g., Liu 2012).

Wavelet analysis can also detect sequence stratigraphic surfaces (Fig. 3.4). Sequence boundary and maximum flooding surfaces may be detected by this method. The wavelet analysis of the wire-line logs can be performed just after logging. So, the results will be available much sooner than the facies analysis of core data. Cores are not available in many cases.

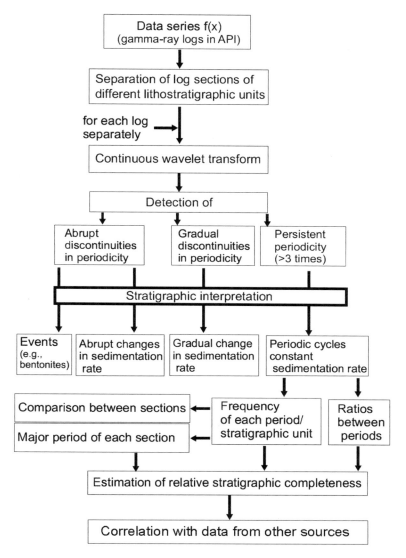

Fig. 3.3 A flow diagram for wavelet analysis of subsurface data (modified from Prokoph and Agterberg 2000)

3.4 Upscaling

Upscaling is the process of assigning an average value to a defined volume. Using upscaling process, different properties can be seen at the various scales of interest. This is useful for comparing and correlation purposes, reservoir modeling and evaluation of heterogeneity at various scales. The averaging methods are different

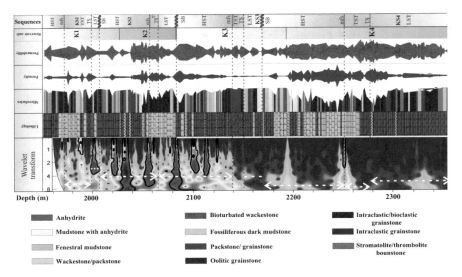

Fig. 3.4 Comparison of continuous wavelet transform of GR data with sequences and facies which have been derived from the geological analysis of the cores in Permian–Triassic reservoirs of the southwest Iran. The Morlet has been used as the standard wavelet

for various properties as well as continuous and discrete logs. Common methods for averaging the discrete logs like facies and diagenetic features include most of, median, minimum, maximum, mid-point pick and arithmetic. Arithmetic, harmonic and geometric means, median, sum, minimum, maximum and mid-point pick are different methods which are applied for upscaling of the continuous properties such as porosity or wire-line logs. Assume a cube of interest which is composed of three different facies (Fig. 3.5). Facies 1 and 3 compose a small part of the cube, while facies 3 fills the most part of it. So, if the averaging method is "most of," the average facies 3 would be the facies of the cube.

In static modeling, upscaling is usually performed in order to assign well-log data to 3D grid cells in well location point. Cell size is an important parameter in determining the final quality and accuracy of the upscaled model. It should be defined based on expected spatial heterogeneity of the reservoir such as lateral and vertical facies distribution. The grid can be refined to fit existing heterogeneity of properties along the well. Upscaled data are compared with the original logs for quality control of upscaling process. If the upscaled logs honor the original data, upscaling has been done correctly. The process is repeated if there is a significant difference between the upscaled and original data. This kind of quality control (QC) can be performed using visual comparison of data or statistical tools such as histograms or cross-plots. To assigning the property into a special facies, bias is used during upscaling. Using this method, only the data corresponding to the facies of interest are used in the averaging process of the cell (Fig. 3.5).

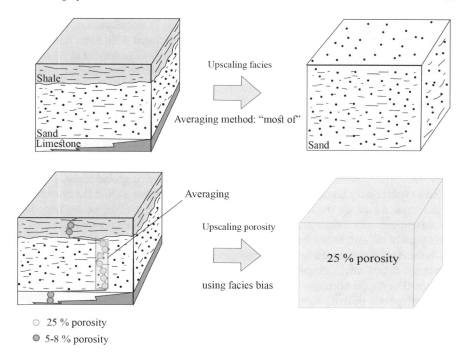

Fig. 3.5   Effect of bias in upscaling the petrophysical properties of a grid cell

3.5   Stratigraphic Correlation

All micro- and mesoscale heterogeneities should be distributed through the basin to understand the properties of interwell spaces. Different mathematical formulas may be used for this purpose, but conceptually, the stratigraphic correlation is a simple method to propagate the reservoir properties to adjacent wells. Stratigraphic layers are traced through the studied field based on the geochronological data. The correlation is based on the principles of stratigraphic order which were established by William Smith (1736–1839) for the first time (Winchester 2001). He defined the principle of superposition and stated that younger strata are deposited on older layers. He also believed that some fossils are characteristic of a certain layer and these layers can be ordered chronologically based on the changes of these fossils through geological time. Based on these principles, strata with the same age can be correlated across various basins. Lithological characteristics as well as fossil content can be used for this purpose. It is worth mentioning that lithological content is not reliable for this purpose but can help in many cases. Different lithologies may have been deposited in the same time, and similar lithologies are deposited in different times (e.g., Tavakoli et al. 2018). At global scale, all strata with the same age can be correlated to each other and a global stratigraphic column is created.

Key stratigraphic surfaces or layers are very useful in correlation of strata in a basin. A regional erosional unconformity can be correlated with high confidence through a field or even a basin. Volcanic ashes or the traces of meteorite in sediments are good examples. They represent an event in a specific time and so are correlated over a large area. At the end of the Cretaceous, a dust layer rich in iridium is detectable worldwide. Such layers or key surfaces are called chronostratigraphic markers. The study of relatively short-lived stratigraphic markers is called event stratigraphy. The term defined by Seilacher (1991) as the sedimentary markers produced by sudden and usually catastrophic short time geological events. These events last for thousands of years or even less. Compared to those which take longer time at geological timescale, these events have been occurred in a short time at regional or global scale.

The evolutionary changes in fossils over geological times enable the correlation of the strata based on the absolute age of sediments. These changes are triggered by abrupt and dramatic environmental variations. The extinction of a species is geologically instantaneous and global in many cases. The planktonic fossils which are floated on water could be correlated globally, but some others are restricted to a specific depth, environment or climate. They are used just for local correlation compared to the first ones.

Geomagnetic polarity timescale is also used for stratigraphic correlations but is limited by the oriented sampling and magnetic minerals which should be oriented by the earth magnetic field during deposition. The geomagnetic polarity is the result of irregular alternations of earth's magnetic field between two different polarity states during Phanerozoic. The geomagnetic polarity remains nearly constant during some time intervals which are called polarity chrons. The present state of the polarity is called normal. So, the opposite state is reverse during the geological time. The profiles are compared with the standard global geometrical polarity, and the strata are correlated (e.g., Zhang et al. 2015; Bover-Arnal et al. 2016).

The seismic data are also used for stratigraphic correlations. These data are provided during exploration phase of the hydrocarbon fields by seismic waves. The formation boundaries and key stratigraphic surfaces are marked on these seismic sections and are correlated through the field or basin (see Sect. 4.1).

The strata can be correlated based on the geochemical characteristics of the carbonate rocks. Chemostratigraphy is determining the chronological order of strata (e.g., elemental composition, stable isotope ratios), arranging them in specific and traceable units and establishing their hierarchy and distribution, based on their geochemical contents (Ramkumar 2015). It is obvious from this definition that chemostratigraphy is a useful method for stratigraphic correlations. Geochemical signatures of the carbonate rocks are changed based on different physical, chemical and biological conditions of the sedimentary environments. At the large scales, these units are homogeneous and so their correlation can help in overcoming the challenges of heterogeneity in carbonate reservoirs. The rocks are characterized based on their geochemical signatures such as elemental concentrations or stable isotope ratios. Sharp boundaries are used for distinguishing the key surfaces. A good example is the global negative excursion of $\delta^{13}C$ in Permian–Triassic boundary (PTB) (e.g., Dolenec et al. 2001; Heydari and Hassanzadeh 2003; Lehrmann et al. 2003;

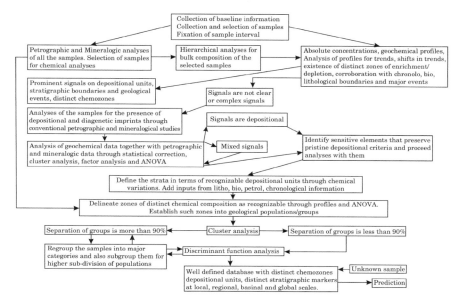

Fig. 3.6 Flowchart for defining chemostratigraphic units and stratigraphic correlation based on these units (from Ramkumar 2015)

Korte et al. 2004a, b; Tavakoli 2015; Tavakoli et al. 2018) which is related to the mass extinction at the end of Paleozoic. The trends of the geochemical changes are also used for chemostratigraphy. The standard sampling procedure, corrections and statistical analysis must be applied for satisfactory results (Fig. 3.6).

References

Al-Dhafeeri AM, Nasr-El-Din HA (2007) Characteristics of high-permeability zones using core analysis, and production logging data. J Petrol Sci Eng 55(1–2):18–36

Amaefule JO, Altunbay MH, Tiab D, Kersey DG, Keelan DK (1993) Enhanced reservoir description using core and log data to identify hydraulic (flow) units and predict permeability in uncored intervals/wells. In: Paper presented at SPE annual technical conference and exhibition, Houston, Texas, 3–6 Oct 1993

Arifianto I, Surjono SS, Erlangga G, Abrar B, Yogapurana E (2018) Application of flow zone indicator and Leverett J-function to characterise carbonate reservoir and calculate precise water saturation in the Kujung formation, North East Java Basin. J Geophys Eng 15(4):1753–1766

Bear J (1972) Dynamics of fluids in porous media. Elsevier, New York

Bover-Arnal T, Moreno-Bedmar JA, Frijia G, Pascual-Cebrian E, Salas R (2016) Chronostratigraphy of the barremian-early albian of the maestrat basin (E Iberian Peninsula): Integrating strontium-isotope stratigraphy and ammonoid biostratigraphy. Newsl Stratigr 49(1):41–68

Dolenec T, Lojen S, Ramovs A (2001) The Permian-Triassic boundary in Western Slovenia (Idri-jca Valley section): magnetostratigraphy, stable isotopes, and elemental variations. Chem Geol 175:175–190

Ebanks WJ (1987) The flow unit concept—an integrated approach to reservoir description for engineering projects. AAPG Meeting Abstracts 1:521–522

Ellwood BB, Tomkin JH, Febo LA, Stuart CN (2008) Time series analysis of magnetic susceptibility variations in deep marine sedimentary rocks: a test using the upper Danian-Lower Selandian proposed GSSP, Spain. Palaeogeogr Palaeocl Palaeoeco 261(3–4):270–279

Eltom H, Makkawi M, Abdullatif O, Alramadan K (2013) High-resolution facies and porosity models of the upper Jurassic Arab-D carbonate reservoir using an outcrop analogue, central Saudi Arabia. Arab J Geosci 6(11):4323–4335

Farshi M, Moussavi-Harami R, Mahboubi A, Khanehbad M, Golafshani T (2019) Reservoir rock typing using integrating geological and petrophysical properties for the Asmari Formation in the Gachsaran oil field, Zagros basin. J Petrol Sci Eng 176:161–171

Flugel E (2010) Microfacies of carbonate rocks, analysis, interpretation and application. Springer, Berlin

Ginsburg RN, James NP (1974) Holocene carbonate sediments of continental shelfs. The geology of continetal margins. Springer, Berlin, pp 137–155

Gunter GW, Finneran JM, Hartman DJ, Miller JD (1997) Early determination of reservoir flow units using an integrated petrophysical method, SPE 38679. In: SSPE annual technical conference and exhibition. San Antonio, TX; October 5–8, 1997

Hear CL, Ebanks WJ, Tye RS and Ranganatha V (1984) Geological Factors Influencing Reservoir Performance of the Hartzog Draw Field, Wyoming. J Petrol Tech, 1335–1344

Herbert TD, Fischer AG (1986) Milankovitch climatic origin of mid-Cretaceous black shale rhythms in central Italy. Nature 321:739–743

Heydari E, Hassanzadeh J (2003) Deevjahi model of the Permian-Triassic boundary mass extinction: a case for gas hydrates as the main cause of biological crisis on Earth. Sediment Geol 163:147–163

Izadi M, Ghalambor A (2013) New approach in permeability and hydraulic-flow unit determination. SPE Reserv Eval Eng 16(3):257–264

Jennings JW, Ruppel SC, Ward WB (2000) Geostatistical analysis of permeability data and modeling of fluid-flow effects in carbonate outcrops. SPEREE 3:292–303

Jensen JL, Lake LW, Corbett PW, Goggin DJ (2000) Statistics for petroleum engineers and geoscientists. Elsevier, Amsterdam, p 338

Korte C, Kozur HW, Joachimski MM, Strauss H, Veizer J, Schwark L (2004a) Carbon, sulfur, oxygen and strontium isotope records, organic geochemistry and biostratigraphy across the Permian/Triassic boundary in Abadeh, Iran. Int J Earth Sci 93:565–581

Korte C, Kozur HW, Mohtat-Aghai P (2004b) Dzhulfian to lowermost Triassic $\delta13C$ record at the Permian/Triassic boundary section at Shahreza, Central Iran. Hallesches Jahrb der Geowissenschaften Beiheft 18:73–78

Lehrmann DJ, Payne JL, Felix SV, Dillett PM, Wang H, Yu Y, Wei J (2003) Permian-Triassic boundary sections from shallow-marine carbonate platforms of the Nanpanjiang Basin, South China: implications for oceanic conditions associated with the end-Permian extinction and its aftermath. Palaios 18:138–152

Lerat O, Van Buchem FSP, Eschard R, Grammer GM, Homewood PW (2000) Facies distribution and control by accommodation within high-frequency cycles of the Upper Ismay interval (Pennsylvanian, Paradox Basin, Utah). In: Homewood PW, Eberli GP (eds) Genetic stratigraphy at the exploration and production scales. Elf EP, Pau, France, pp 71–91

Liu Z (2012) Orbital cycles analysis and its genesis significance for the sequence hierarchy: a case study of Carboniferous Karashayi Formation, Central Tarim basin. J Earth Sci 23(4):516–528

Liu Y, Liu Y, Zhang Q, Li C, Feng Y, Wang Y, Xue Y, Ma H (2019) Petrophysical static rock typing for carbonate reservoirs based on mercury injection capillary pressure curves using principal component analysis. J Petrol Sci Eng 181:106175

Mirzaei-Paiaman A, Saboorian-Jooybari H, Pourafshary P (2015) Improved method to identify hydraulic flow units for reservoir characterization. Energy Technol 3:726–733

Mirzaei-Paiaman A, Ostadhassan M, Rezaee R, Saboorian-Jooybari H, Chen Z (2018) A new approach in petrophysical rock typing. J Petr Sci Eng 166:445–464

Mirzaei-Paiaman A, Sabbagh F, Ostadhassan M, Shafiei A, Rezaee MR, Saboorian-Jooybari H, Che Z (2019) A further verification of FZI and PSRTI: newly developed petrophysical rock typing indices. J Petr Sci Eng 175:693–705

Nabawy BS, Rashed MA, Mansour AS, Afify WSM (2018) Petrophysical and microfacies analysis as a tool for reservoir rock typing and modeling: Rudeis Formation, off-shore October Oil Field, Sinai. Mar Petrol Geol 97:260–276

Nazari MH, Tavakoli V, Rahimpour-Bonab H, Sharifi-Yazdi M (2019) Investigation of factors influencing geological heterogeneity in tight gas carbonates, Permian reservoir of the Persian Gulf. J Petrol Sci Eng 183, art. no 106341

Prokoph A, Agterberg FP (2000) Wavelet analysis of well-logging data from oil source rock, Egret Member, Offshore Eastern Canada. AAPG Bull 84(10):1617–1632

Prokoph A, Thurow J (2000) Diachronous pattern of Milankovitch cyclicity in late Albian pelagic marlstones of the North German Basin. Sed Geol 134(3–4):287–303

Prokoph A, Shields GA, Veizer J (2008) Compilation and time-series analysis of a marine carbonate $\delta^{18}O$, $\delta^{13}C$, $^{87}Sr/^{86}Sr$ and $\delta^{34}S$ database through earth history. Earth-Sci Rev 87(3–4):113–133

Rahimpour-Bonab H, Mehrabi H, Navidtalab A, Izadi-Mazidi E (2012) Flow unit distribution and reservoir modelling in cretaceous carbonates of the Sarvak Formation, Abteymour Oilfield, Dezful Embayment, SW Iran. J Petr Geol 35(3):213–236

Ramkumar M (2015) Chemostratigraphy: concepts, techniques and applications. Elsevier, Amsterdam

Read JF (1982) Carbonate platforms of passive (extensional) continental margins: types, characteristics and evolution. Tectonophysics 81:195–212

Riazi Z (2018) Application of integrated rock typing and flow units identification methods for an Iranian carbonate reservoir. J Petr Sci Eng 160:483–497

Schwarzacher W (1993) Cyclostratigraphy and Milankovitch theory. Devel Sedimentol 52:225

Seilacher A (1991) Events and their signatures: an overview. In: Einsele G, Ricken W, Seilacher A (eds) Cycles and events in stratigraphy. Springer, Berlin, Heidelberg, New York, p 955

Shaw AB (1964) Time in stratigraphy, 365. McGraw-Hill Book Company, New York

Soleymanzadeh A, Parvin S, Kord S (2019) Effect of overburden pressure on determination of reservoir rock types using RQI/FZI, FZI and Winland methods in carbonate rocks. Petr Sci. https://doi.org/10.1007/s12182-019-0332-8

Tavakoli V (2015) Chemostratigraphy of the Permian-Triassic Strata of the Offshore Persian Gulf, Iran. In: Ramkumar M (ed) Chemostratigraphy: concepts, techniques, and applications. Elsevier, Amsterdam

Tavakoli V (2018) Geological core analysis: application to reservoir characterization. Springer, Cham

Tavakoli V, Rahimpour-Bonab H, Esrafili-Dizaji B (2011) Diagenetic controlled reservoir quality of South Pars gas field, an integrated approach. C R Geosci 343:55–71

Tavakoli V, Naderi-Khujin M, Seyedmehdi Z (2018) The end-Permian regression in the western Tethys: sedimentological and geochemical evidence from offshore the Persian Gulf, Iran. Geo-Mar Lett 38(2):179–192

Tiab D, Donaldson EC (2015) Petrophysics, theory and practice of measuring reservoir rock and fluid transport properties. Gulf Professional Publishing, Houston

Walker RG (1984) Facies models. Geological Association of Canada

Winchester S (2001) The map that changed the World. Harper Collins, New York, p 329

Zhang Y, Li M, Ogg JG, Montgomery P, Huang C, Chen ZQ, Shi Z, Enos P, Lehrmann DJ (2015) Cycle-calibrated magnetostratigraphy of middle Carnian from South China: implications for Late Triassic time scale and termination of the Yangtze Platform. Palaeogeo Palaeocl Palaeoeco 436:135–166

Chapter 4
Macroscopic Heterogeneity

Abstract Microscopic and mesoscopic heterogeneities should be combined to form larger-scale units which can be correlated across the field or basin. This is performed by using seismic data, sedimentary strata, sequence stratigraphic concepts, constructing various maps and reservoir tomography. Seismic waves pass through the reservoir rocks and illustrate the large-scale structures. These data are very useful for detailed subsurface correlations at field scale. Layers with acoustic impedance contrast represent different zones. They should be carefully interpreted. These seismic lines may indicate formation tops, chronological boundaries, different lithologies or any change in reservoir properties. Stratification is caused by changes in facies properties through time. They represent different depositional conditions and probably reservoir characteristics. The widespread lateral continuity of the strata makes them useful for large-scale correlations. Genetically related strata are classified as sequence stratigraphic units. These units are deposited in the same physical, chemical and biological conditions, and so, they usually have similar reservoir properties. With respect to their primary textures, they also have the same diagenetic imprints. The boundaries between these units are marked based on the core analysis data, seismic traces and wire-line logs. They are correlated between the wells as reservoir zones. Then, microscopic and mesoscopic scale units are distributed in sequence stratigraphic framework. Low permeability or impermeable layers or strata are considered as static or dynamic compartmentalizers in the reservoir body. They strongly affect the fluid properties of reservoir compartments and so should be carefully considered before production. Data from various wells are projected on horizontal planes as maps or on vertical sections as profiles. They help to improve the macroscopic understanding of the reservoir heterogeneities. The horizontal maps of different depths are used for reservoir tomography. The variations in properties through time are revealed with this method.

4.1   Seismic Interpretation

Seismic data are derived from the artificially induced seismic events. The acoustic waves are sent out into the earth where various structures, geological and non-geological objects reflect these waves. The reflection from any object depends on its acoustic impedance which is the product of velocity and density. The reflected data are processed to illustrate a visual representation of the earth interior. Both 2 and 3D sections are produced and used for further interpretations. Generally, the resolution of the seismic sections is not higher than tens of meters and so cannot be used for identifying the heterogeneity at microscales, but these reflectors are good indicators of bedding planes. Two sides of a reflection have different acoustic impedance. The non-geological reflections such as diffractions or multiples must be removed first. Non-sedimentary features such as faults or fluid contacts also produce reflections which can be used for further interpretations.

The basic rule is that the main reflections represent the bedding planes such as formation tops. Most of these beddings are chronological boundaries and so are correlated through the field or basin (Fig. 4.1).

These reflections are depicted as large-scale images, and so lateral variations are clearly observed. These boundaries indicate a major change in depositional environments such as energy level, sedimentation rate, type of carbonate platforms or major diagenetic impacts on the reservoir. The reflection continuity indicates the lateral variability of the layers or formations. The shallow and deep sediments are correlated based on a continuous reflection. This is very important because usually the effects of depositional changes in deep environments are not clear. For example, a sequence boundary is clearly defined based on the change in stacking pattern in shallow settings but there is no clear evidence of sea-level changes in basinal

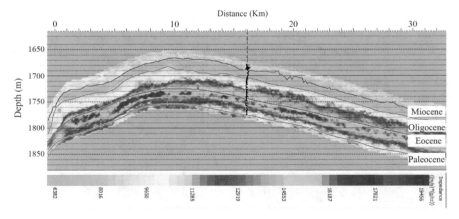

Fig. 4.1   An example of a seismic section from southwest Iran showing the impedance of the strata. Chronological boundaries are obvious

environments. Reflection amplitude is the result of contrast between two geological layers or formations. It may also be the result of different fluid content. The amplitude changes regarding the contrast in lithology, porosity or fluid content. The geometry of the geological layers is deduced from reflection configuration. It shed light on the depositional and tectonic processes which formed the primary geological structures such as paleotopography or syndepositional faulting of the basement. Reflection frequency is used for calculating the bed thickness and possible fluid content. The interval velocity is derived from the seismic reflections. This velocity indicates porosity, pore type, fluid content and lithology of the formations. Seismic reflections are also used for distinguishing unconformities and truncations. Two sides of these surfaces have generally high contrast in lithology or fluid content as well as different slopes of the beddings and so can be easily distinguished on seismic sections.

The main application of seismic sections is sequence stratigraphy of the studied strata. As mentioned, erosional surfaces and truncations are recognizable on these sections. They are used for determining the genetically related strata and various systems tracts (STs). In fact, the first appearance of sequence stratigraphy was based on the interpretation of seismic data (Vail et al. 1977). Sequence stratigraphy is used for correlation of the strata at larger scales and reservoir zonation (Sect. 4.4).

Similar reflections are grouped together, and a seismic facies is formed (Sangree and Widmier 1977). The definition is very similar to the basic concept of facies. A set of seismic responses with similar patterns is defined as a seismic facies (Fig. 4.2). Similarity is interpreted based on the reflection configuration, continuity, amplitude

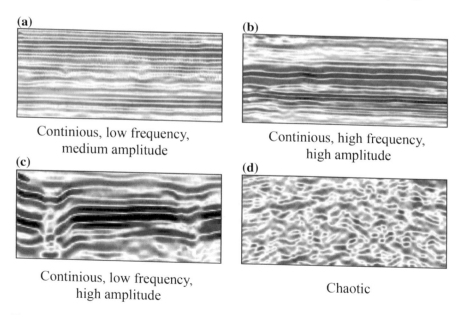

(a) Continious, low frequency, medium amplitude

(b) Continious, high frequency, high amplitude

(c) Continious, low frequency, high amplitude

(d) Chaotic

Fig. 4.2 Examples of seismic facies in sedimentary basins

and frequency (Veeken and van Moerkerken 2013). The seismic facies are described in terms of lithology, genetic conditions and depositional environments. Like other facies types, there are some features which should be considered about each facies. The organization of the internal reflections should be known. In terms of geometry, reflections may be parallel or subparallel, wavy, divergent, hummocky or chaotic (Veeken and van Moerkerken 2013). External geometry is the 3D shape of the overall seismic facies (Vail et al. 1977). Like facies models, each geometry indicates a distinct depositional environment. Examples are sheet-like geometry which shows a calm and widespread depositional setting. Wedges are formed when a sudden break occurs in sedimentation. Good examples are forced regressive wedges in basinal settings which are eroded sediments from the shallower depths. The mound geometry in seismic facies is usually seen because of the presence of a buildup in carbonate platform. Slope fronts are easily recognized and indicate a rimmed shelf or distally steepened ramps or generally a sharp slope in the environment.

Seismic data acquisition, processing and interpretation play an important role in evaluation of reservoir heterogeneity. In fact, seismic sections are images of subsurface at macroscale. The geometry and spatial distribution of facies are easily recognizable with these sections. Lack of seismic data can lead to erroneous interpretations of the reservoir heterogeneity (Fig. 4.3).

4.2 Fracturing

Carbonate fractured reservoirs are important hydrocarbon-producing formations in many giant fields (such as Gachsaran Oil Field in Iran or Kirkuk in Iraq). Various types of fractures may be developed in a reservoir which significantly increases the heterogeneity of the formations (Fig. 4.4).

Fractures have direct impact on permeability by creating connected paths for fluids (Fig. 4.5a). They also affect the porosity by moving the fluid through the rocks and creating empty spaces, such as vugs (Fig. 4.5b). In spite of such importance, fractures are ignored in many reservoir studies. This is because of the general lack of quantitative data about fractures and also the complex interaction between matrix porosity, permeability and fractures (Nelson 2001). Providing a fractured sample (core) is difficult. Measuring their petrophysical properties such as porosity and permeability is not simple. In fact, fractured samples are ignored in fluid related core analysis tests such as relative permeability or overburden pressure experiments. Even with good preservation, cores are not oriented in many cases. Image logs such as formation micro imager (FMI) or oil-base microimager (OBMI) are frequently used for identification, analysis and interpretation of fractures. These are orientated logging tools, and so fracture strike and dip direction may be deduced from their collected data. Cores can be oriented using correlation of strike and dip direction of recognizable core fractures with these data from image logs. At larger scales, seismic sections are used to identify the major faults. Fractures are classified experimentally (shear, extension, tensile) or based on their natural occurrence (tectonic, regional, contractional

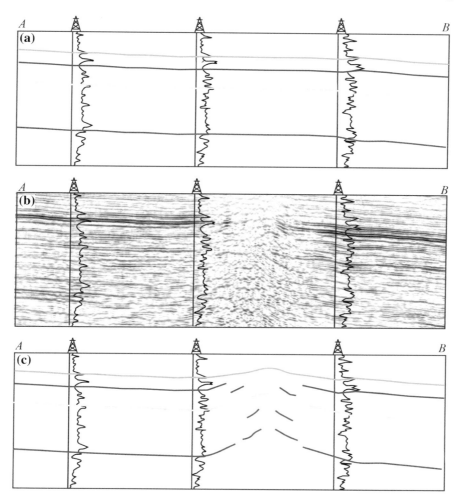

Fig. 4.3 Correlation of three wells in a field. Some seismic anomalies are not clear in log-based correlation (a) while they are recognizable using seismic data (b, c)

and surface-related) (Nelson 2001). Contractional fractures have negligible effect on reservoir properties (Fig. 4.5c–e). To understand the effect of fractures on reservoir quality and heterogeneity, the petrophysical properties of the fractures' network must be addressed. These properties include fracture type, dimension, and morphology, stratigraphically and spatially fracture distribution, geological characteristics (such as cementation or variations due to preferential lithology), structural evolution and petrophysical properties resulting from fracture system of the rocks. Petrophysical properties include permeability and porosity of the fracture, fluid saturation and the recovery factor expected from the fractures of the reservoir (Nelson 2001). In case of permeability, fractures considerably change the anisotropy and heterogeneity of

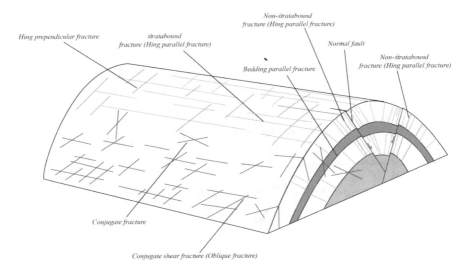

Hing prependicular fracture   *stratabound fracture (Hing parallel fracture)*   *Non-stratabound fracture (Hing parallel fracture)*   *Normal fault*   *Non-stratabound fracture (Hing parallel fracture)*   *Bedding parallel fracture*   *Conjugate fracture*   *Conjugate shear fracture (Oblique fracture)*

Fig. 4.4 Different types of fractures are developed in a reservoir. They have strong effect on fluid flow and heterogeneity of the carbonates (courtesy of R. Nozaem). The figure has been compiled from Hatcher (1994) and Davis et al. (2011)

the rocks (e.g., Hennings et al. 2012; Guerriero et al. 2013; Wang et al. 2016; Volatili et al. 2019).

In terms of morphology, both open and mineral-filled fractures change the heterogeneity of the reservoir. While open fractures increase the permeability of the reservoir, mineral filled fracture acts as fluid barriers and may produce a compartmentalizer (Fig. 4.5f) layer (e.g., Massaro et al. 2018).

All fractures are secondary in origin and so may follow the other formation properties. The effect of lithology and facies characteristics on fracturing has been discussed by many authors (e.g., Moore and Wade 2013; Michie 2015; Dong et al. 2017; Korneva et al. 2018). These studies indicate that lithology, crystal size, texture, bed thickness and porosity type, size and volume have strong effects on fracturing. Many researchers believe that fracturing is more intense in dolomites compared to limestone (e.g., Schmoker et al. 1985; Nelson 2001; Gale et al. 2004; Korneva et al. 2018). Some studies have documented the opposite (e.g., Beliveau et al. 1993; Rustichelli et al. 2015). Flugel (2010) and Rustichelli et al. (2015) demonstrated that fine crystalline dolostones are densely fractured compared to coarse-crystalline dolomites. Sowers (1970) and McQuillan (1973) showed that the fracture spacing increases with increasing the bed thickness. So, fractures follow the precursor heterogeneity of the rocks in some cases.

Fracture analysis of outcrop sections shed light on the fracture development in carbonate reservoirs (e.g., De Keijzer et al. 2007; Bisdom et al. 2017; Burberry and Peppers 2017; Massaro et al. 2018). Outcrops are available over large areas, the geometry and thickness of the layers can be traced continuously, and fractures are

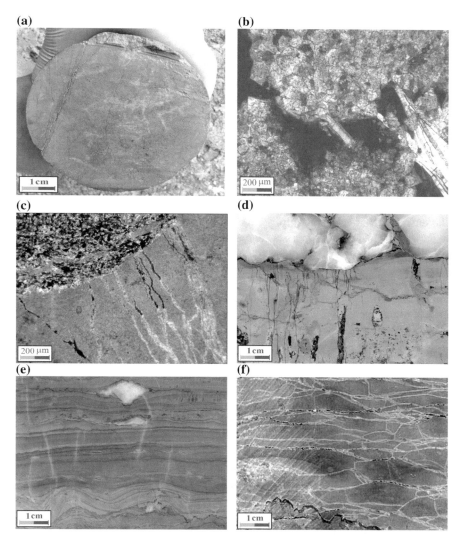

Fig. 4.5 A highly fractured core which has a high potential for fluid flow (Oligo–Miocene Asmari Formation) (**a**), fractures enlarge the fluid pathways and create the vugs (Oligo–Miocene Asmari Formation) (**b**), small fractures are formed by anhydrite dehydration (Triassic Kangan Formation) (**c, d**), contractional fractures in intertidal stylolites (Triassic Kangan Formation) (**e**) and a dense network of fractures act as a compartmentalizer layer (Permian Dalan Formation) (**f**). **b** and **c** cross-polarized light. Others are core samples

clearly seen. Types and dimensions of the faults and fractures are obvious. The stratigraphic and 3D fracture heterogeneity as well as evolutionary patterns of the fractures can be visually identified on the rocks. Care should be taken when comparing the results. Fractures and their resulting permeability are very sensitive to overburden pressures and fluid content, which are not present at surface conditions. Layers and strata may be laterally heterogeneous, and so geological and petrophysical properties can be completely different between surface and subsurface formations. The effect of weathering also should be considered.

Fracture stratigraphy is used for organizing the fractures within similar mechanically layered sequences (e.g., Morettini et al. 2005; De Keijzer et al. 2007; Ferrill and Morris 2008; Ameen et al. 2009; Dashti et al. 2018; Lavenu and Lamarche 2018). The rock properties such as lithology, facies, bedding thickness and layer interfaces as well as tectonic activities in reservoirs control the spatial distribution of fractures. The term "mechanical stratigraphy" is used to subdivide the reservoir rocks based on their geomechanical properties such as elastic stiffness and tensile strength. Distribution of these units is a good indicator of the reservoir heterogeneity in terms of fracturing and its propagation. Fractures' properties such as frequency, spacing, morphology and orientation are characterized by cores, image logs or seismic sections. It is obvious that these units, such as other types of reservoir heterogeneities, are scale-dependent.

4.3 Stratification

The term "stratification" has originated from the word stratum (single) and strata (plural). It means formation of different rock layers in various sedimentary environments. Strata represent special environmental conditions in the past. A set of strata are called a zone which is limited by the chronostratigraphic boundaries. Originally, stratal surfaces may be horizontal or inclined based on the topography of the substrate. Most of the sedimentary rocks are stratified at the time of deposition. The thickness of the strata may be different across the basin. It can be changed from cm to km size. Changes in texture (e.g., size, shape and even color of grains) as well as composition of the sediments cause stratification. Other variabilities might occur in layer dip, thickness and form, which might result in rather complex heterogeneities. Variation in properties causes stratification, and so sediments and rocks within a layer are homogeneous at larger scales. The lateral heterogeneities within a layer are considerable in some cases and should be considered. In carbonate depositional environments, facies types indicate the grain to matrix ratio, grain sizes and compositions. Depending on depositional settings, facies are different and consequently stratified carbonates are deposited. Progradation or retrogradation of carbonate deposits also causes stratification of the sediments. Beds with different facies are stacked on top of each other. For example, deep mudstone facies with planktonic foraminifera are precipitated on previously deposited high-energy shoal facies through the sea-level flooding. Depending on depositional environments, the shape and size of strata may be different. Alternation of thin strata increases the vertical heterogeneity of the formation.

In subsurface sections, study of stratification and stratified intervals is performed based on the subsurface data such as cores, well logs and seismic sections. Cores are more reliable, and the strata can be visually recognized and differentiated. Although cores have some limitations because of the small size (usually 4 inch in diameter), they provide direct data from the stratified layers, their thickness, shape and variation in composition and textures. Correlation of strata based on core analysis results is helpful in modeling the petrophysical behavior of the reservoir (e.g., Borgomano et al. 2008) or understanding their lateral variations. Beside the cores, well logs and especially different kinds of image logs provide valuable data on the stratification in the subsurface. Modern tools have high resolution. They give information on thickness of beds, their strike, dip, dip direction, lithology (composition), facies (texture) and the shape of beddings. Cross-stratifications are developed in carbonates, especially in grain-dominated facies. Ooid grainstones are the best example. They are also detectable by core samples or image logs.

Seismic data are useful when their resolution is high and the strata are thick enough to be viewed by these sections. Seismic data are gathered at the field scale (tens to many tens of kilometers), and so they are good source of information about the lateral extension and the stacking patterns of deposits. This kind of data also gives clear indication of strata terminations.

4.4 Sequence Stratigraphy

For many years, the main criteria for dividing a formation into reservoir units were a challenging problem. Lithostratigraphy and biostratigraphy were the bases of reservoir zonation for many years. The main problem was variable reservoir properties within the same lithologies or one biozone. The boundary of lithostratigraphic units is determined based on lithological characteristics, but various lithologies may have similar reservoir properties (Fig. 4.6). In contrast, similar lithologies may have different reservoir behavior. This is also the case for biostratigraphy. The boundaries between units are determined chronologically. They indicate the timelines, and so the biostratigraphical units have the same reservoir properties in many cases, but this is not necessarily true (Fig. 4.6). So, petroleum geologists used the genetically related strata for reservoir zonation and the sequence stratigraphy was born. These units have had similar physical, chemical and biological conditions at the time of deposition, and so they should have similar reservoir properties. Sequence stratigraphic-based reservoir zonation defines homogeneous reservoir units at larger scales, compared to micro- and mesoscale units such as core-scale rock types or electrofacies. These units are bounded by major fluctuations in sea level at the time of deposition. There are various models to interpret these changes. Applying a model to the studied basin or field depends on the sedimentological properties and reservoir characteristics of the strata (Catuneanu et al. 2009).

(a)

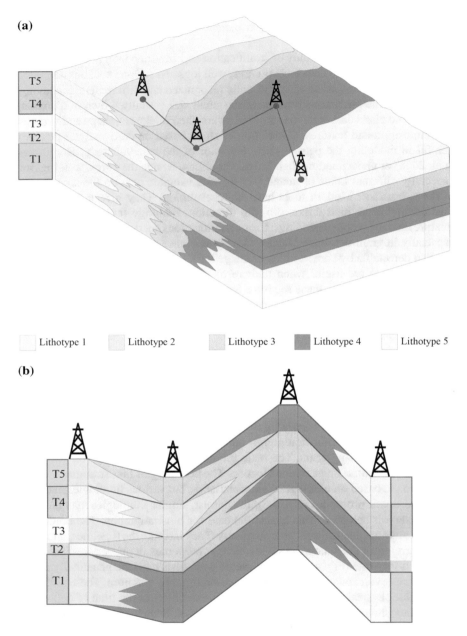

| | Lithotype 1 | | Lithotype 2 | | Lithotype 3 | | Lithotype 4 | | Lithotype 5 |

(b)

Fig. 4.6 Lithostratigraphic and chronological boundaries do not coincide with each other in some cases. The 3D model (**a**) and fence diagram (**b**) from a hypothetical basin show this concept. Lithotypes 1 and 2 are reservoir units

Diagenesis may change the rock properties after deposition, but as the primary characteristics of these strata are similar, their diagenetic alterations are also equal. For example, meteoric waters dissolve the aragonite rather than calcite. In ooid grainstones, ooids are aragonitic and so the moldic porosities follow the primary facies. Early diagenesis occurs very close to the time of deposition and so it selectively affects the depositional textures. Some diagenetic processes such as fracturing are non-fabric-selective anyway. Even these diagenetic processes follow the precursor heterogeneity of the samples in some cases (see Sect. 4.2). Figure 4.7 shows the idealized sequences of Permian–Triassic carbonate reservoirs of the Persian Gulf. The porosity–permeability cross-plot of these sequences and their STs are illustrated in Fig. 4.8. It can be seen that there is a good relationship between these two parameters in some cases. The reservoir quality of each unit is also different from the others.

Constructing the porosity–permeability relationships for understanding the heterogeneity of the formation is the first step. The results of evaluating heterogeneity at various scales should be integrated with each other to better understand the causes of variations and the best methods to overcome this challenge. As mentioned before (Sect. 2.8), rock typing is the final step in microscale. Rock types and flow units are defined using petrophysical parameters, but geological rock types (GRTs) help

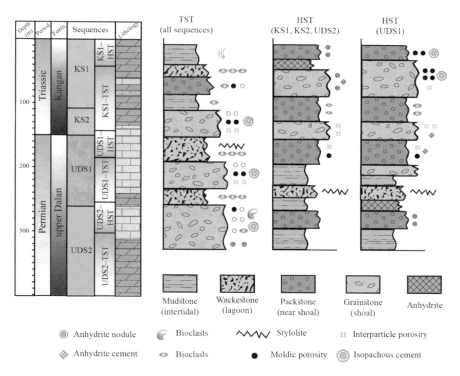

Fig. 4.7 Idealized facies stacking of each ST in Permian–Triassic formations of the Persian Gulf. HST: highstand systems tract, TST: transgressive systems tract

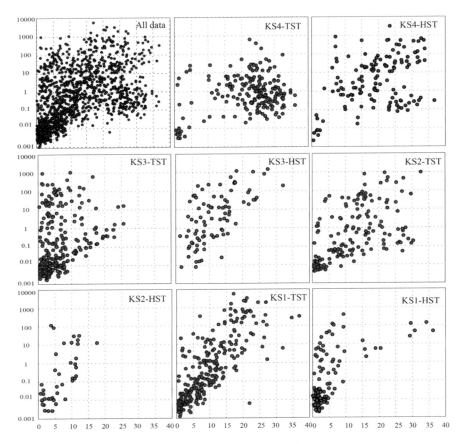

Fig. 4.8 Porosity–permeability distribution within STs. Sedimentological and stratigraphical properties of the sequences can be seen from Fig. 4.7

to understand the effective primary as well as secondary factors which control the heterogeneity of the formation. So, GRTs, flow zone indicator, Loernz plot, Winland rock typing and other methods should be applied to obtain an accurate characterization of reservoir heterogeneity. Then, their distribution in various ST is determined. Due to the high heterogeneity of carbonates, there are always some uncertainties. The main objective is to define a homogeneous unit, as much as possible.

4.5 Reservoir Compartmentalization

Reservoir compartmentalization occurs when a petroleum accumulation is divided into different parts (compartments) from fluid type or pressure point of view (Jolley et al. 2010). These segments behave as separate flow units during production because compartmentalizers prevent fluid flow through the reservoir units. Well-known examples are cap rocks which are capable of trapping the hydrocarbon over geological times. These are "static compartmentalizers" (Jolley et al. 2010) and are routinely characterized at the exploration and appraisal stages of the value chain of the reservoir. There are also some seals which allow the fluid to flow across the reservoirs in long geological times but strongly reduce the flow in the time of production. These seals have very low permeability and are called "dynamic compartmentalizers" as their role changes after appraisal of the reservoir. At the first stages of exploration and appraisal, the formation pressure data approximate a straight line according to the pressure gradient in the reservoir. After production, the fluids and pressures are not allowed to equilibrate between reservoir layers and so the reservoirs are segregated into different fluid/pressure zones. This is a serious problem because field developments are planned according to the well production of a specific region or formation. Unexpected compartmentalization led to the unexpected changes in reservoir pressure in various units and geographic areas. It also causes internal blowout which is uncontrolled flow from high-pressure zones into the low-pressure units. Compartmentalization also changes the position and geometry of the contacts. So, the recoverable hydrocarbon from drilled wells may change considerably. The behavior of the stress-related phenomena such as breakouts or fractures also changes along with pressure anomalies. The phase changes of fluids are also possible. All these events are resulted from reservoir heterogeneity which should be evaluated before production.

Several studies of both carbonate and siliciclastic reservoirs have shown that reservoir compartmentalization is a common phenomenon which is controlled by a variety of geological factors such as sedimentology, diagenesis, stratigraphy and/or tectonism at various scales (e.g., Clark et al. 1996; Rahimpour-Bonab 2007; Jolley et al. 2010; Rahimpour-Bonab et al. 2014; Mehrabi et al. 2016; Hosseini et al. 2018; Nabawy et al. 2018). Mud-dominated carbonates have low porosity and permeability in most cases. They are deposited in intertidal, lagoon and basin depositional environments. In carbonate–evaporite settings, evaporative minerals such as gypsum and anhydrite are precipitated as interparticle cements as well as separate layers. Gypsum is dehydrated in burial environment and transforms to anhydrite which reduces reservoir quality, increases heterogeneity and acts as compartmentalizer layer. Sheet-like geometry of these deposits makes them a good candidate for separating the reservoir units. Mudstones and wackestones of the lagoon environment texturally have high potential to be compartmentalizers, but their lens-shaped geometry is not appropriate for preventing fluid flow across the layers. Basin mud-dominated facies also have sheet-like shape and widespread development, but their facies associations are not suitable for hydrocarbon. Diagenetic processes also may create a low porosity and permeability interlayer. High rate of cementation occludes the porosities and reduces the permeabilities of the rocks and makes a dynamic compartmentalizer. A good example is the lower Triassic rocks of the Zagros Basin in Iran. After the Permian–Triassic mass extinction, the seawater was supersaturated with respect to carbonates because the carbonate secreting organisms were extinct. So, thrombolites were expanded in the early Triassic epiric sea of the Arabian Plate. Thrombolites are made from fine micrite particles with low permeability (Abdolmaleki and Tavakoli 2016; Abdolmaleki et al. 2016). High rate of cementation also occluded the rare pore spaces, and so a dynamic compartmentalizer was created (Fig. 4.9). This layer with very low permeability allowed the hydrocarbon to flow across the Permian–Triassic boundary but acted as a seal at the time of production. So, the strata which seemed to be one reservoir, segregated into two reservoirs with different pressure regime.

Vertical sealed faults and fractures may compartmentalize the different geographic regions of a field (e.g., Bailey et al. 2002; Ainsworth 2006; Massaro et al. 2018). So, the field may be both horizontally and vertically compartmentalized. The fractures or faults must be impermeable to segregate the reservoir into different regions. This is very important in shared fields where production from one side in one country

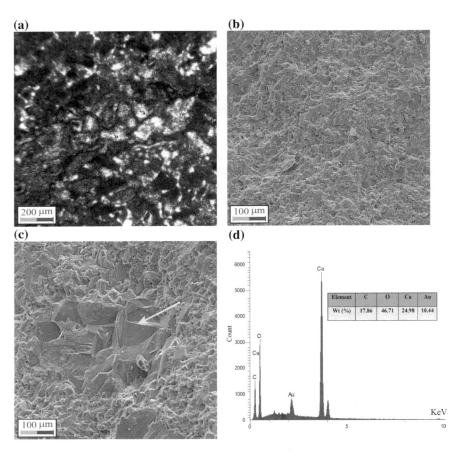

Fig. 4.9 Post-extinction thrombolite facies of the lower Triassic strata of Zagros Basin in thin section (**a**) and SEM (**b**) images. Large calcite crystals have been precipitated from supersaturated seawater (**c**). Energy-dispersive X-ray (EDX) analysis confirms the calcite mineralogy (**d**)

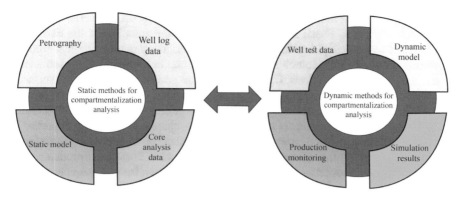

Fig. 4.10 Reservoir compartmentalization is studied based on various static and dynamic data and is monitored during the lifetime of the reservoir

strongly affects the pressure and fluid properties in another country. Such heterogeneities may completely change the financial and development plans of the field. A comprehensive study on reservoir compartmentalization and understanding its causes and consequences is necessary for any hydrocarbon field during appraisal stage of development. After production, the results should be verified using dynamic production data (Fig. 4.10).

4.6 Reservoir Zonation

At macroscopic scale, reservoir zonation is the most important step to overcome the challenge of reservoir heterogeneity. So, many researchers tried to introduce an effective method for dividing the reservoir interval into homogeneous units. In contrast to the microscopic scale in which one sample is considered at a time and so there is no intrinsic heterogeneity, different samples are included in a reservoir zone. So, defining a homogeneous unit at macroscopic scale is not a simple procedure. The goal is to combine the similar adjacent samples into one reservoir-scale zone, as much as possible. The boundaries should be determined carefully and traceable throughout the studied field. Many researchers tried to introduce such a procedure based on various methods of zonation (e.g., Martin et al. 1997; Daraei et al. 2017; Tavakoli 2017; Jodeyri-Agaii et al. 2018).

It seems that the lithofacies differentiation is the simplest form of zonation. These boundaries are easily recognized on cores, cuttings, thin sections and wire-line logs. There is always a forecast for lithology variations in new wells. Petroleum geologists and drillers have used this method of zonation for many years. In spite of its advantages, lithological changes are not representative of the reservoir properties in many cases. Limestones have a wide range of porosity values, permeabilities, pore throat size distribution, wettability and so on. This is also the case about dolomites.

So, the application of lithological zonation is limited to new drilled wells, especially in less known fields. Fossil contents are also used for zonation of the carbonate reservoirs. Chronological boundaries are determined using paleontological studies. These boundaries usually coincide with reservoir zones because age-related strata are deposited in similar conditions. Unconformities are good examples of time surfaces in reservoirs (e.g., Tavakoli et al. 2018). They control the reservoir characteristics of their lower and upper strata (e.g., Abdolmaleki et al. 2016). This criterion is not applicable when index fossils are not present. Also, change in reservoir properties is not related to time in some cases. Instead, lateral changes in depositional environments strongly affect the reservoir potential of the rocks. These environments with different reservoir properties are deposited simultaneously. Instead, sequence stratigraphic surfaces are defined based on the genetic relationships between strata. These sediments have been deposited in similar physical, geochemical and biological conditions which cause their same reservoir properties. As mentioned earlier, diagenetic processes, especially early diagenesis, are usually similar in these sedimentary rocks. So, the accepted method for reservoir zonation is sequence stratigraphy. The most important aspect is that many sequence stratigraphic concepts are model-dependent (Catuneanu et al. 2009; Catuneanu 2019) and so an appropriate model should be selected to fulfill the changes in reservoir properties. Anyway, model-independent sequence stratigraphic aspects are used as the standard criteria in various geological settings, scales and for different available data. At first, stratal stacking patterns are recognized and the framework of STs is defined. These are model-independent and so can be defined in any basin or field. Then, sequence boundaries are selected and specific types of STs are delineated. The last two, however, are model-dependent and should be selected according to the conditions of the case study such as available type and resolution of data, depositional environments, tectonics and paleoclimate. For example, the forced regressive systems tract (FRST) is not recognizable in arid carbonate ramps in many cases. Defined microscale rock types and flow units are then distributed in STs and the degree of heterogeneity is determined. This can be started from a simple porosity–permeability cross-plot in one zone (ST or sequence) and continue with distributing geological rock types (GRTs) in these units. The frequency of core-scale rock types in each sequence stratigraphic unit is also considered which indicate the degree of heterogeneity in a zone. Other statistical parameters such as coefficient of variation or suitable charts such as pie diagrams or histograms are applied in order to evaluate the heterogeneity of each unit. It is obvious that any defined zone must be traceable, mappable and correlatable through the field or even the basin. This, in turn, can achieve using the wire-line log data. Defined boundaries are correlated with the peaks or trough of the various logs. Wire-log log characteristics of each zone are extracted and correlated across the field or basin. Electrofacies (EF) (Sect. 2.9) are also used for such correlations. Like rock types, distribution of EFs is also considered in each zone and they are traced in other wells with no cores.

Vertical distribution of flow units can also indicate the reservoir zones. Cumulative plot of flow units versus depth shows the reservoir boundaries. In a perfect equality situation when there is no vertical heterogeneity in the reservoir, the increasing rate with depth is constant. So, plotting depth versus cumulative flow zones gives a straight line with constant slope. Considerable changes in this slope indicate major variations in flow units and consequently reservoir properties. The normal probability plot (NPP) of flow zones also yields the same result. This plot illustrates the normal scores versus the data set and a diagonal straight line indicates a normal distribution. Normal data (normal scores) distribution is symmetric, unimodal and the mean, median and mode are all equal. There are built-in commands for creating a NPP in mathematical software such as Minitab. They plot the sorted data against the quantiles. Quantiles are subdivisions of a data set into equal parts and so have a normal distribution. For example, you can divide a distribution into 100 equal parts and call it percentiles. These percentiles divide the data into quantiles with 100 equal proportions. Quantiles are calculated by inverse of the standard normal cumulative distribution which has a mean of zero and a standard deviation of one. Different slopes in this line indicate different zones (see Fig. 3.2a).

Reservoir zonation is also performed using stratigraphic modified Lorenz plot. Originally, Lorenz technique was developed to measure the degree of inequality in wealth across a population (Lorenz 1905). The method is modified by petroleum engineers through plotting cumulative flow capacity versus cumulative thickness of the sampling intervals (Schmalz and Rahme 1950). In a homogeneous reservoir, the property is constantly increasing with depth. So, the result of plotting cumulative values against depth will be a diagonal straight line which is known as the "line of perfect equality." With increasing heterogeneity, the property will change randomly and this causes a departure from the line of perfect equality. The new line is Lorenz curve, and the area between the Lorenz curve and the line of perfect equality multiply by two is called Lorenz coefficient (Lc). With increasing heterogeneity, the Lc also increases from zero (maximum homogeneity) to one (maximum heterogeneity). Different reservoir-scale flow units can be assessed by stratigraphic modified Lorenz plot (SMLP) (Gunter et al. 1997). All data are arranged in depth order. Each porosity value is multiplied by the thickness of sampling interval and the cumulative frequency is calculated. Results are normalized to 1 or 100%. This is called storage capacity. The process is repeated for permeability data, and the result is flow capacity. The SMLP is constructed by plotting storage capacity versus flow capacity. Reservoir zones are differentiated based on the changes in slope of the resulting line (Fig. 4.11). The mentioned methods such as porosity–permeability plot or core-scale rock type distribution are used for validating the determined zones (Fig. 4.11), as discussed previously.

4.7 Maps

The main goal of heterogeneity evaluation in carbonate reservoirs is reconstruction of the interwell space with high resolution and accuracy. Many developed methods are based on analyzing vertical data sets (such as rock types, flow units and Lorenz plot) which are applied to each well separately. Some others have been introduced to

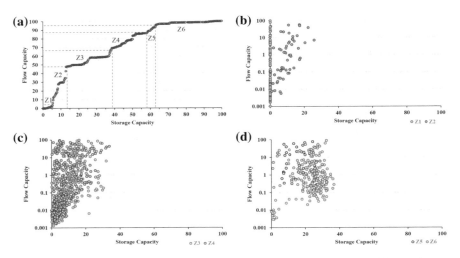

Fig. 4.11 SMLP and the resulting zones for the Oligo–Miocene carbonate strata of the south–west Zagros Basin (**a**). Each subfigure shows porosity–permeability data distribution in two zones (**b–d**). According to these results, zone 3 can be divided into two new zones

correlate data qualitatively (such as facies models). Maps are quantitative interpolation of interwell reservoir properties within geological frameworks. Routinely, they are 2-dimensional representation of the property distribution on a plane in the subsurface. The accuracy and resolution of this distribution depend on the type and amount of available data and used methods of interpolation. These data comes from cores, wire-line logs or seismic sections. With increasing data, the accuracy and resolution of the subsurface visualizations are also improved. In fact, maps are 2D representation of data before building a 3D model and reconstructing the spatial distribution of reservoir properties. They help to decide about further investment and drilling. Any data set with longitudinal and latitudinal coordination as well as depth position can be illustrated as a map. These include elevation, porosity, permeability, water saturation, lithology, facies and rock types, to name a few. For preparing a subsurface map, it is essential to use accurate data, consider all possible interpretations and use the most suitable mapping techniques to achieve a realistic map which is consistent with the geological reality of the field. So, property interpolation just based on the mathematical equations is not recommended in carbonate reservoirs. They are highly heterogeneous, and properties change according to geological variables, not statistical basis. Instead, discrete property values should be attributed to specific facies. Then, their distribution is constructed based on the concepts of facies models. An understanding of the structural framework as well as local structures of the field is necessary before any interpretation.

While all maps are not reconstructed by contouring techniques, most of them use this method to representing data on a 2D surface. The main parts of a contour map are isolines. They are lines of equal vales. Various data are contoured in different maps. For example, points with the same stratigraphic thickness can be connected to each other and an isopach map is constructed. A proper contour interval is selected based on the range of data. There are some basic rules that must be followed during contouring. For example, contours cannot cross each other. This is obvious because a point has just one value. Closer contours indicate higher gradient and vice versa. Routinely, data are contoured by computers because they are very fast and accurate. The user chooses a proper contouring algorithm to achieve a realistic map. Ideally, a grid with regularly spaced X, Y and Z is made. Then, the grid nods (grid line intersects) are calculated based on the available data and the final contour map is constructed based on this grid. The spacing and so the number of nodes are determined by user based on the real data spacing.

Various types of subsurface maps are utilized to understand the heterogeneity of carbonate reservoirs. The most commonly used is structural contour map. This map marks a key horizon or surface in terms of its elevation regarding a horizontal datum which is usually mean sea level. The points are routinely obtained by drilling. The structural contour map of formation top is useful for future drilling programs and also understanding the reservoir geometry. An isopach map shows the true stratigraphic thickness of an individual unit. The contours are called isopach lines which indicate the same thickness of the unit of interest. Widely spaced lines mean slight lateral thickness variations and closely spaced lines represent rapid changes in thickness. These types of maps are utilized to understand the geometry and volume of the reservoir rocks, recognizing the stratigraphic traps, identifying depositional environments and positioning new wells for exploration and production. An isochore map illustrates the vertical thickness of a unit, layer or formation.

Facies maps show the variations in facies. They are very useful to evaluate the heterogeneity and understanding the distribution of properties in the reservoir because particular facies commonly have their unique reservoir characteristics. The lateral extent and variations in reservoir and non-reservoir zones can be identified with these maps. A facies map can be generated just based on gross lithologies, but for more sophisticated exploration purposes, it is necessary to create a detailed facies map to accurately locate the hydrocarbon-bearing intervals. Various types of facies maps include:

- Lithology maps which illustrate the lithology distribution.
- Isolith maps show the thickness of single lithology.
- Percentage map which the contours indicate the percentage of a single unit in a desired thickness.
- Ratio maps that are generated based on the ratio of two lithologies. The mud-dominated/grain-dominated ratio is a good example in carbonate reservoirs.
- Triangle facies maps are used for depicting the ratio of three lithologies. Actually, these are not contour maps. The ratios of various lithologies or facies are illustrated using the patterns and shadings. Each lithology or facies type lies at one apex of a triangle, and their ratios are characterized by different patterns (Fig. 4.12). Each data point in the reservoir has a known ratio of three facies. These values are connected to each other, and the final map is constructed (Fig. 4.12).

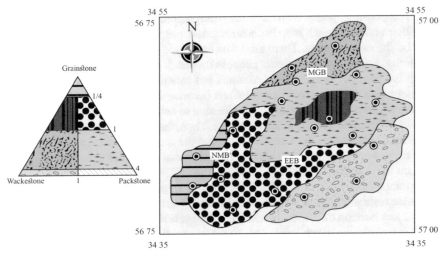

MGB: Mud-dominated and grain-dominated boundary

NMB: No mud boundary   EEB: Equilibrium energy boundary

⊙   Well location

Fig. 4.12 A triangle map of a Jurassic carbonate formation in northwest Iran

A paleogeographic map represents the geographic position of the geological characteristics in the past. Some types of these maps are useful for understanding the reservoir heterogeneity. For example, they may represent the distribution of various facies or lithologies in different geographic positions in a specific geological time. This helps to understand the type of sedimentary environments and reservoir rocks' distribution. In these maps, the topmost deposits are removed and the geological map of the area at the time of deposition is illustrated.

In a reservoir with no major structural and tectonic elements, drawing the maps follows the basic rules of contouring but main faults can change the rules. The main problem is that detecting subsurface faults is difficult. They are not adequately revealed by contouring. Normal faults result in missing stratigraphic intervals in the wells, while reverse faults repeat the intervals. Both missing and repeated sections may be revealed by comparing wire-line logs from at least two wells (Fig. 4.13).

Major faults are also detectable on seismic sections. Surface mapping and field observations are helpful in some cases.

Sections and profiles are 2D correlations of wells. The interwell space is projected onto a vertical plane, and so the properties can be traced along a specific direction (Fig. 4.14). Fence diagrams are 2D sections which are viewed in a 3D space. They are very useful for comparing the distribution of the properties in various directions (Fig. 4.14).

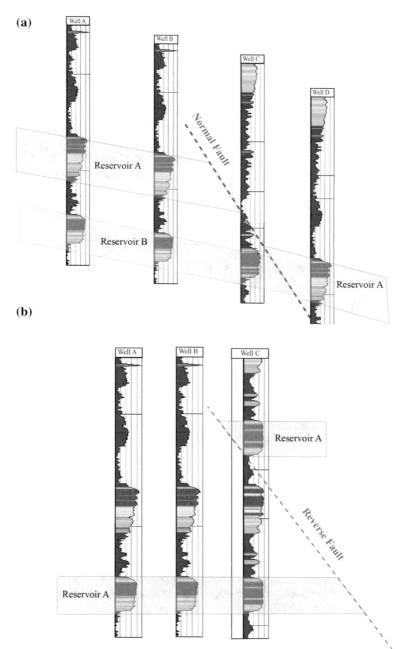

Fig. 4.13 Both normal (**a**) and reverse (**b**) faulting can be revealed by comparing wire-line log data from at least two wells

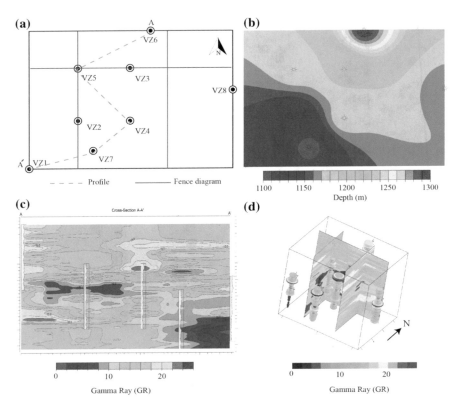

Fig. 4.14 Various maps for the top of the Albian carbonate sequence in west Iran. The location of the wells (**a**), elevation of the formation tops (**b**), a profile (**c**) and a fence diagram (**d**) are obvious. The traces of the profile and fence are visible on the location map

4.8   Reservoir Tomography

Reservoir tomography is generating horizontal slices through the interpolation of data from different wells of a reservoir. It is a newly developed technique in reservoir characterization. The term "tomography" in reservoir studies generally points to the computerized tomography (CT) by X-ray method, which discussed previously (Sect. 2.3). Seismic tomography illustrates a 3D picture of the earth's interior using seismic waves. As previously discussed, this is a macroscopic tool for understanding the reservoir heterogeneity (Sect. 4.1). Here, tomography means serial horizontal sectioning of the reservoir with sequential maps. Same types of data from various wells are needed to construct such serial sections, and so core-derived properties or wire-line logs are used for this purpose. Horizontal distribution of reservoir properties and vertical changes through time are easily recognized with these maps. It is also possible to make an animation which clearly shows the changes in reservoir properties through time (horizontal slices). Comparing various tomographic images, the relationships between various reservoir parameters in macroscopic scale can be analyzed. Tomographic maps are also used to reconstruct the depositional environments and the primary facies distribution in the studied reservoir (e.g., Wylie and Huntoon 2003; Wylie and Wood 2005). They are also used for understanding the diagenetic processes, their causes and consequences by interpolating the thin sections' data or wire-line log readings. For example, reservoir tomography of the pore types in a carbonate reservoir with moldic (dissolution) and intercrystalline (dolomitization) porosities indicates the evolution of the fluid–rock interactions in the reservoir. Both vertical and horizontal wells can be planned using these maps. Routinely, slices are illustrated in a stacked column (Fig. 4.15). For example, Wylie and Wood (2005) used the GR, porosity and permeability tomography to reconstruct the detailed anatomy of the pinnacle reef of Silurian Brown Niagara reservoir in the Belle River Mills field (Fig. 4.16). They concluded that there is a strong correlation between porosity, permeability and GR values in this reservoir.

Fig. 4.15 Stacked GR slices of an Albian carbonate ramp in west Iran. Changes in GR values through time could be due to the terrestrial input or change in oxidation state of the depositional environment

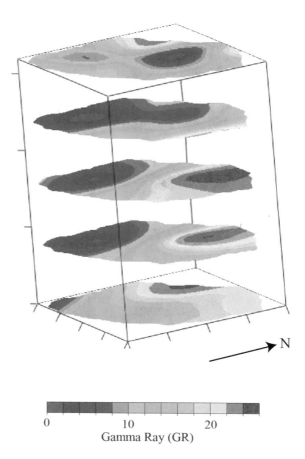

Gamma Ray (GR)

0   10   20

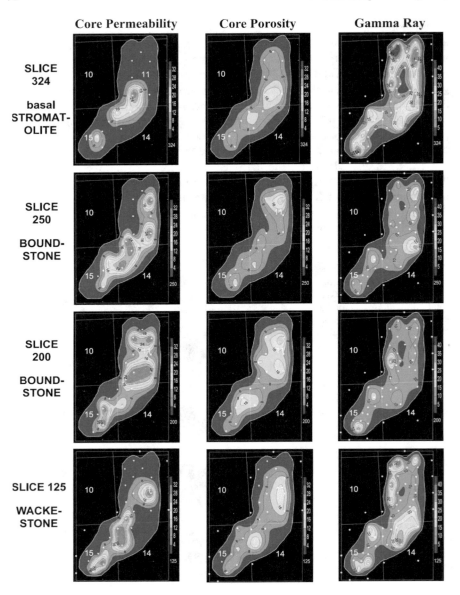

Fig. 4.16 Comparison of four porosity, permeability and GR slices for Silurian Brown Niagara reservoir (Wylie and Wood 2005). The same trend is observed in all three parameters. Slices have been numbered from bottom to top, and so 125 is at the base and 324 lies at the top of the pinnacle reef. Permeability is in mD, porosity is in percent, and GR is in API. Whit dots are control points

4.9 Macroscopic Uncertainties

Uncertainty in macroscopic heterogeneity is strongly affected by uncertainties in data gathering. The low resolution of seismic data compared with core plugs or wire-line logs does not allow for continuous records of properties. The experience of the analyst and his/her geological knowledge about the studied area have strong effect on the final interpretations. The situation is more complicated for fractures. Many difficulties exist in their data acquisition, petrophysical evaluation and characterization. The problem is direct sampling from the reservoirs which is difficult for fracture study. Their petrophysical properties are difficult to determine because most part of the fluids flow through the fractures in laboratory tests. Quantitative approaches for fracture analysis in subsurface studies have not been developed. A deterministic approach is necessary for understanding the fluid flow in a fractured media which is not available yet (Nelson 2001). Image logs and seismic data can help to overcome these difficulties, but image logs are not available from all boreholes and seismic data have low resolution for fracture imaging.

There are also some uncertainties in sequence stratigraphy and reservoir zonation. Various models are used for determining sequences and STs in a carbonate reservoir. Selecting an appropriate model and positioning the boundaries depend on the geologist's experience. There are different criteria to distinguish a proper zone to justify the fluid storage and flow. In many cases, the reliability of the work is determined by distribution of microscopic results such as rock types within each zone. These microscopic units, in turn, have some uncertainties which make the problem more complicated. Tracing the zones and mapping their distribution is another problem. This could be unrealistic without accurate seismic data or a comprehensive geological background of the field. The accuracy of a map also depends on the used algorithm which may change the result. Anyway, with all of these uncertainties, considerable care must be taken when predicting the spatial distribution of properties in a carbonate reservoir.

References

Abdolmaleki J, Tavakoli V (2016) Anachronistic facies in the early Triassic successions of the Persian Gulf and its palaeoenvironmental reconstruction. Palaeogeo Palaeoclimat Palaeoeco 446:213–224

Abdolmaleki J, Tavakoli V, Asadi-Eskandar A (2016) Sedimentological and diagenetic controls on reservoir properties in the Permian-Triassic successions of Western Persian Gulf, Southern Iran. J Petr Sci Eng 141:90–113

Ainsworth RB (2006) Sequence stratigraphic–based analysis of reservoir connectivity: influence of sealing faults—a case study from a marginal marine depositional setting. Petr Geosci 12:127–141

Ameen MS, Smart BGD, Somerville JMc, Hammilton S, Naji NA (2009) Predicting rock mechanical properties of carbonates from wireline logs (A case study: Arab–D reservoir, Ghawar field, Saudi Arabia). Mar Petr Geol 26(4):430–444

Bailey WR, Manzocchi T et al (2002) The effect of faults on the 3D connectivity of reservoir bodies: a case study from the East Pennine Coalfield, UK. Petr Geosci 8:263–277

Beliveau D, Payne DA, Mundry M (1993) Waterflood and CO_2 flood of the fractured midale field. J Petr Technol 45(9):817–881

Bisdom K, Nick HM, Bertotti G (2017) An integrated workflow for stress and flow modelling using outcrop–derived discrete fracture networks. Comput Geosci 103:21–35

Borgomano JRF, Fournier F, Viseur S, Rijkels L (2008) Stratigraphic well correlations for 3–D static modeling of carbonate reservoirs. AAPG Bull 92(6):789–824

Burberry CM, Peppers MH (2017) Fracture characterization in tight carbonates: an example from the Ozark Plateau, Arkansas. AAPG Bull 101(10):1675–1696

Catuneanu O (2019) Model–independent sequence stratigraphy. Earth-Sci Rev 188:312–388

Catuneanu O, Abreu V, Bhattacharya JP, Blum MD, Dalrymple RW, Eriksson PG, Fielding CR, Fisher WL, Galloway WE, Gibling MR, Giles KA, Holbrook JM, Jordan R, Kendall CGStC, Macurda B, Martinsen OJ, Miall AD, Neal JE, Nummedal D, Pomar L, Posamentier HW, Pratt BR, Sarg JF, Shanley KW, Steel RJ, Strasser A, Tucker ME, Winker C (2009) Towards the standardization of sequence stratigraphy. Earth–Sci Rev 92(1–2):1–33

Clark MS, Beckley LM, Crebs TJ, Singleton MT (1996) Tectono–eustatic controls on reservoir compartmentalisation and quality—an example from the upper miocene of the San Joaquin basin, California. Mar Petr Geol 13(5):475–491

Daraei M, Bayet-Goll A, Ansari M (2017) An integrated reservoir zonation in sequence stratigraphic framework: A case from the Dezful Embayment, Zagros, Iran. J Petr Sci Eng 154:389–404

Dashti R, Rahimpour-Bonab H, Zeinali M (2018) Fracture and mechanical stratigraphy in naturally fractured carbonate reservoirs–a case study from Zagros region. Mar Petr Geol 97:466–479

Davis GH, Reynolds SJ, Kluth CF (2011) Structural geology of rocks and regions. Wiley, Hoboken, NJ

De Keijzer M, Hillgartner H, Al Dhahab S, Rawnsley K (2007) A surface–subsurface study of reservoir–scale fracture heterogeneities in Cretaceous carbonates. N Om Geol Soc Spec Publ 270:227–244

Dong T, Harris NB, Ayranci K, Yang S (2017) The impact of rock composition on geomechanical properties of a shale formation: middle and Upper Devonian Horn River Group shale, Northeast British Columbia, Canada. AAPG Bull 101(2):177–204

Ferrill DA, Morris AP (2008) Fault zone deformation controlled by carbonate mechanical stratigraphy, Balcones fault system, Texas. AAPG Bull 92(3):359–380

Flugel E (2010) Microfacies of carbonate rocks. Analysis, interpretation and application, 2nd ed, p 984 XXIII

Gale JFW, Laubach SE, Marrett RA, Olson JE, Holder J, Reed RM (2004) Predicting and characterizing fractures in dolostone reservoirs: using the link between diagenesis and fracturing. In: Braithwaite CJR, Rizzi G, Darke G (eds) The geometry and petrogenesis of dolomite hydrocarbon reservoirs, vol 235. Geological Society, London (Special Publications), pp 177–192

Guerriero V, Mazzoli S, Iannace A, Vitale S, Carravetta A, Strauss CA (2013) permeability model for naturally fractured carbonate reservoirs. Mar Petr Geol 40(1):115–134

Gunter GW, Finneran JM, Hartmann DJ, Miller JD (1997) Early determination of reservoir flow units using an integrated petrophysical method. In: Proceedings—SPE annual technical conference and exhibition, Omega (Pt 1), pp 373–380

Hatcher RD (1994) Structural geology: principles concepts and problems. Prentice Hall, Upper Saddle River

Hennings P, Allwardt P, Paul P, Zahm C, Reid R Jr, Alley H, Kirschner R, Lee B, Hough E (2012) Relationship between fractures, fault zones, stress, and reservoir productivity in the Suban gas field, Sumatra, Indonesia. AAPG Bull 96(4):753–772

Hosseini M, Tavakoli V, Nazemi M (2018) The effect of heterogeneity on NMR derived capillary pressure curves, case study of Dariyan tight carbonate reservoir in the central Persian Gulf. J Petr Sci Eng 171:1113–1122

Jodeyri-Agaii R, Rahimpour-Bonab H, Tavakoli V, Kadkhodaie-Ilkhchi R, Yousefpour MR (2018) Integrated approach for zonation of a mid–cenomanian carbonate reservoir in a sequence stratigraphic framework. Geol Acta 16(3):321–337

Jolley SJ, Fisher QJ, Ainsworth RB (2010) Reservoir compartmentalization: an introduction. In: Jolley SJ, Fisher QJ, Ainsworth RB, Vrolijk PJ, Delisle S (eds) Reservoir compartmentalization. Geological Society, London (Special Publications), p 347

Korneva I, Bastesen E, Corlett H, Eker A, Hirani J, Hollis C, Gawthorpe RL, Rotevatn A, Taylor R (2018) The effects of dolomitization on petrophysical properties and fracture distribution within rift related carbonates (Hammam Faraun Fault Block, Suez Rift, Egypt). J Struct Geol 108:108–120

Lavenu APC, Lamarche J (2018) What controls diffuse fractures in platform carbonates? Insights from Provence (France) and Apulia (Italy). J Struct Geol 108:94–107

Lorenz MO (1905) Methods of measuring concentration of wealth. Am Stat Assoc 9(70):209–219

Martin AJ, Solomon ST, Hartmann DJ (1997) Characterization of petrophysical flow units in carbonate reservoirs. AAPG Bull 81(5):734–759

Massaro L, Corradetti A, Vinci F, Tavani S, Iannace A, Parente M, Mazzoli S (2018) Multiscale fracture analysis in a reservoir–scale carbonate platform exposure (Sorrento Peninsula, Italy): implications for fluid flow. Geofluids Art No. 7526425

McQuillan H (1973) Small-scale fracture density in Asmari formation of Southwest Iran and its relation to bed thickness and structural setting. AAPG Bull 47(12):2367–2385

Mehrabi H, Mansouri M, Rahimpour-Bonab H, Tavakoli V, Hassanzadeh M, Eshraghi H, Naderi M (2016) Chemical compaction features as potential barriers in the Permian-Triassic reservoirs of South Pars Field, Southern Iran. J Petr Sci Eng 145:95–113

Michie EAH (2015) Influence of host lithofacies on fault rock variation in carbonate fault zones: a case study from the Island of Malta. J Struct Geol 76:61–79

Moore CH, Wade WJ (2013) Natural fracturing in carbonate reservoirs. Dev Sedimentol 67:285–300

Morettini E, Thompson A, Eberli G, Rawnsley K, Roeterdink R, Asyee W, Christman P, Cortis A, Foster K, Hitchings V, Kolkman W, van Konijnenburg JH (2005) Combining high–resolution sequence stratigraphy and mechanical stratigraphy for improved reservoir characterisation in the Fahud field of Oman. GeoArabia 10(3):17–44

Nabawy BS, Basal AMK, Sarhan MA, Safa MG (2018) Reservoir zonation, rock typing and compartmentalization of the Tortonian-Serravallian sequence, Temsah Gas Field, offshore Nile Delta, Egypt. Mar Petr Geol 92:609–631

Nelson RA (2001) Geologic analysis of fractured reservoirs, 2nd ed, p 352

Rahimpour-Bonab H (2007) A procedure for appraisal of a hydrocarbon reservoir continuity and quantification of its heterogeneity. J Petr Sci Eng 58(1–2):1–12

Rahimpour-Bonab H, Enayati-Bidgoli AH, Navidtalab A, Mehrabi H (2014) Appraisal of intra reservoir barriers in the Permo-Triassic successions of the central Persian gulf, offshore Iran. Geol Acta 12(1):87–107

Rustichelli A, Iannace A, Girundo M (2015) Dolomitization impact on fracture density in pelagic carbonates: contrasting case studies from the Gargano Promontory and the southern Apennines (Italy). Ital J Geosci 134(3):556–575

Sangree JB, Widmier JM (1977) Seismic stratigraphy and global changes in sealevel Part 9: seismic interpretation of clastic depositional facies. In: Payton (ed) Seismic stratigraphy: application to hydrocarbon exploration, vol 26. AAPG Memoir, pp 165–184

Schmalz JP, Rahme HS (1950) The variations in water flood performance with variation in permeability profile. Prod Mon 15(9):9–12

Schmoker JW, Krystinik KB, Halley RB (1985) Selected characteristics of limestone and dolomite reservoirs in the United States. AAPG Bull 69(5):733–741

Sowers GM (1970) Private report "Theory of spacing of extension fractures," in geologic fractures of rapid excavation. Geol Soc Am Eng, Geol. Case History no 9:27–53

Tavakoli V (2017) Application of gamma deviation log (GDL) in sequence stratigraphy of carbonate strata, an example from offshore Persian Gulf, Iran. J Petr Sci Eng 156:868–876

Tavakoli V, Naderi–Khujin M, Seyedmehdi Z (2018) The end–Permian regression in the western Tethys: sedimentological and geochemical evidence from offshore the Persian Gulf, Iran. Geo–Mar Lett 38(2):179–192

Vail PR, Todd RG, Sangree JB (1977) Seismic stratigraphy and global changes of sea level: part 5. Chronostratigraphic significance of seismic reflections: section 2. Appl Seism Reflect Config Strat Interpret 26:99–116

Veeken PCH, van Moerkerken B (2013) Seismic stratigraphy and depositional facies models. EAGE Publications

Volatili T, Zambrano M, Cilona A, Huisman BAH, Rustichelli A, Giorgioni M, Vittori S, Tondi E (2019) From fracture analysis to flow simulations in fractured carbonates: The case study of the Roman Valley Quarry (Majella Mountain, Italy). Mar Petr Geol 100:95–110

Wang F, Li Y, Tang X, Chen J, Gao W (2016) Petrophysical properties analysis of a carbonate reservoir with natural fractures and vugs using X–ray computed tomography. J Nat Gas Sci Eng 28:215–225

Wylie AS Jr, Huntoon JE (2003) Log curve amplitude slicing—visualization of log data for the Devonian Traverse Group, Michigan basin, U.S. AAPG Bull 87(4):581–608

Wylie AS Jr, Wood JR (2005) Well–log tomography and 3–D imaging of core and log–curve amplitudes in a Niagaran reef, Belle River Mills field, St. Clair County, Michigan, United States. AAPG Bull 89(4):409–433

Chapter 5
Petrophysical Evaluations

Abstract Formation evaluation is one of the most important parts of a reservoir study. This process is mainly performed using wire-line logs. These data are usually available from all wells and reservoir intervals. Various formulas and parameters are used for calculating porosity, water saturation and lithology from logs. Reservoir heterogeneities may considerably change these formulas and their parameters. Matrix responses, pore types, permeabilities and other petrophysical properties are different for each unit or rock type. Therefore, applying one formula for all samples yields erroneous results. These results are used for calculating the final amount of oil in place. The results of these calculations change future development plans and investments in the field. After partitioning the reservoir into different rock types or reservoir units, the appropriate parameters are used in the relevant calculations. These include the log responses for various lithologies of different units as well as special petrophysical characters (such as Archie's cementation factor) for the unit. In deterministic approach, these are different for each step. In other words, unique parameters are used for each formula in different units or rock types. Then, the results of the previous stage are used for the next calculation. In a stochastic approach, different mathematical models are developed for calculating the reservoir properties in each homogeneous unit. Parameters are different for each model. For example, if pore types control the distribution of water saturation in a reservoir, the reservoir is classified into various parts according to its pore types. Then, a separate model is developed for each part and laboratory-derived data for each rock type are combined with wire-line logs for calculating the water saturation of the entire reservoir interval.

5.1 Effects of Heterogeneity

Books have been written on theories, concepts and applications of petrophysics in reservoir studies (e.g., Lucia 2007; Tiab and Donaldson 2015). The purpose of this chapter is considering the effects of heterogeneity on petrophysical evaluations and how the problem may be solved. Petrophysics is the study of fluid flow and storage. The properties of interest such as porosity, permeability and water saturation are mainly derived from core analysis (McPhee et al. 2015; Tavakoli 2018). However,

© The Author(s), under exclusive license to Springer Nature Switzerland AG 2020

V. Tavakoli, *Carbonate Reservoir Heterogeneity*,
SpringerBriefs in Petroleum Geoscience & Engineering,
https://doi.org/10.1007/978-3-030-34773-4_5

cores are not available from most wells, while wire-line log data are gathered from almost all wells in a field. So, the rock and fluid properties are calibrated to wire-line logs and correlated to the other parts of the field. For many years, these wire-line log data have been used for calculating petrophysical properties of the reservoir rocks. So, usually the term "petrophysics" is used for evaluating the petrophysical properties based on these data. This concept is used in this chapter.

The major goal of wire-line logging is calculating three properties including porosity, lithology and water saturation. Various formulas or mathematical algorithms may be applied to the log data, and different results are gained. However, the main purposes are still constant. Many equations have been developed to calculate the mentioned properties from these data. Some of the formulas are experimental (e.g., Archie equation for calculating water saturation), and some others are derived from petrophysical concepts using mathematical rules (such as Wyllie formula for calculating porosity from sonic velocity). The heterogeneity of rocks has major effects on the results of both groups. So, the final plan for field development and investments may vary considerably. High heterogeneity of the rocks causes erroneous results. All of these formulas have constant values or assume special situations which are not the same for all case studies. In fact, they are strongly different for various samples. For example, the Archie's first law (Eq. 5.1) states that cementation exponent (m) is a function of resistivity of formation water (R_w), bulk resistivity of a rock fully saturated with fluid of resistivity R_w (R_o) and porosity as:

$$R_o = R_w \Phi^{-m} \tag{5.1}$$

The m is used for calculating water saturation in hydrocarbon reservoirs. For a group of samples, m is calculated by plotting formation factor (F) versus porosity on a log–log scatter plot (Fig. 5.1). F is calculated by dividing R_o to R_w.

Archie states that m varies with respect to different parameters. These parameters determine the tortuous pathways for fluid flow. For a direct path, such as fractures, the m is 1 but in carbonates with complicated pore structures, m varies between 2 and 2.6. So, grouping the rocks based on various properties such as their pore types, pore throat sizes and ranges of permeabilities can yield different results for water saturation in the samples (Figs. 5.2, 5.3; Tables 5.1, 5.2 and 5.3). These calculations were done in Permian–Triassic carbonate reservoirs of the Persian Gulf basin (Nazemi et al. 2018). The water saturation of the rocks is then calculated using the Archie's formula (Eq. 5.2) as follows:

$$S_w = \sqrt[n]{\frac{R_w \Phi^{-m}}{R_t}} \tag{5.2}$$

where S_w is water saturation, Φ is porosity, and R_t is true resistivity of the sample. The n is saturation exponent which is assumed the usual value of 2 in this example to avoid more complexity.

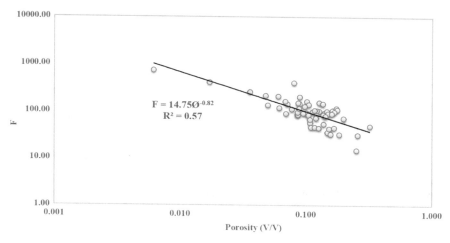

Fig. 5.1 Plot of porosity versus formation factor for carbonate samples of the Permian–Triassic reservoir of the Persian Gulf. The slope of the line is $-m$ (Nazemi et al. 2018)

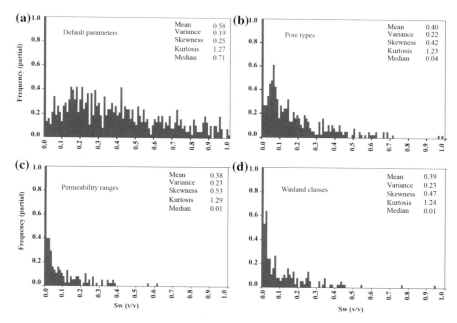

Fig. 5.2 Difference between calculated S_w with different grouping methods in a carbonate formation. S_w of the samples has been calculated with the default values (2 for both m and n) for all samples (**a**). Then, it has been recalculated in samples with different pore types (**b**), permeability ranges (**c**) and pore throat sizes (**d**), separately. The Archie exponents have been calculated in laboratory for each group

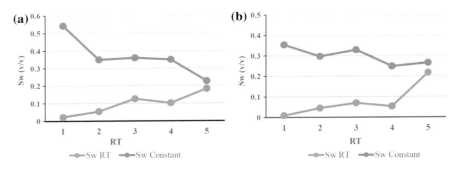

Fig. 5.3 S_w calculation of a tight carbonate formation in Persian Gulf basin. The difference between calculated S_w with constant Archie's exponents (orange line) and variable exponents (blue line) is obvious. The resulted S_w in two wells (**a**, **b**) is higher with constant values. Rock types (RT) have been determined based on the ranges of permeabilities

Table 5.1 Statistical parameters and m exponents in Permian–Triassic reservoirs of the Persian Gulf based on classifying pore types

Pore type	m	R^2	m (mean)	m (SD)	m (CV)	Equations
Intercrystalline	0.76	0.91	1.75	0.32	0.18	$F = 14.94\varnothing^{-0.76}$
Interparticle	1.74	0.56	2.11	0.22	0.10	$F = 2.18\varnothing^{-1.74}$
Moldic	0.76	0.57	2.15	0.39	0.18	$F = 17.67\varnothing^{-0.76}$
Vuggy	1.74	0.99	1.74	0.04	0.02	$F = 0.99\varnothing^{-1.71}$

R^2 has been calculated between F and porosity. SD standard deviation, CV coefficient of variation (modified from Nazemi et al. 2018)

Table 5.2 Statistical parameters and m exponents in Permian–Triassic reservoirs of the Persian Gulf based on various pore throat sizes

R_{35} (μm)	m	R^2	m (mean)	m (SD)	m (CV)	Equations
R_{35} < 0.2	0.77	0.92	2.07	0.27	0.13	$F = 19.80\varnothing^{-0.77}$
R_{35} 0.2–0.5	0.7	0.65	2.05	0.34	0.17	$F = 19.72\varnothing^{-0.70}$
R_{35} 0.5–1	0.62	0.19	2.16	0.44	0.20	$F = 23.19\varnothing^{-0.62}$
R_{35} 1–2	0.86	0.82	2.06	0.40	0.19	$F = 12.54\varnothing^{-0.86}$
R_{35} 2–5	2.45	0.91	2.11	0.16	0.07	$F = 0.50\varnothing^{-2.45}$

R^2 has been calculated between F and porosity. SD standard deviation, CV coefficient of variation, R_{35} pore throat sizes in 35% mercury saturation in mercury injection capillary pressure (MICP) test (modified from Nazemi et al. 2018)

It is shown in Tables 5.1, 5.2, 5.3 and Fig. 5.2 that the m of the groups and resulted S_w are completely different.

The same goes for the formulas that have some constants. These constants are completely different for various rock types. A good example is calculating the pore

Table 5.3 Statistical parameters and m exponents in Permian–Triassic reservoirs of the Persian Gulf based on various ranges of permeability

Permeability (mD)	m	R^2	m (mean)	m (SD)	m (CV)	Equations
0.01–0.1	0.30	0.98	1.92	0.4	0.21	$F = 53.74\varnothing^{-0.30}$
0.1–1	0.73	0.62	2.03	0.31	0.15	$F = 19.11\varnothing^{-0.73}$
1–10	0.92	0.31	2.23	0.40	0.18	$F = 12.64\varnothing^{-0.92}$
10–100	1.12	0.89	1.91	0.31	0.16	$F = 4.94\varnothing^{-1.12}$

R^2 has been calculated between F and porosity. SD standard deviation, CV coefficient of variation (modified from Nazemi et al. 2018)

throat sizes using Winland equation. Winland's equation relates the porosity, permeability and pore throat sizes in 35% saturation in MICP test as follows (Kolodize 1980) (Eq. 5.3):

$$\log(R_{35}) = 0.732 + 0.588 \log(K_{air}) - 0.864 \log(\Phi) \tag{5.3}$$

where R_{35} is pore throat sizes in 35% mercury saturation, K_{air} is air permeability, and Φ is porosity. It is obvious that these constants have been calculated using the Winland's database which was 56 sandstones and 26 carbonate samples. These constants should be modified for each formation and its rock types to increase the accuracy of the formula for prediction pore throat sizes or permeabilities. In spite of such importance, few studies have modified these coefficients (e.g., Nooruddin et al. 2016) and most of them used the original ones. Nooruddin et al. (2016), based on MICP analysis from multiple Jurassic aged carbonate reservoirs in the Middle East region, concluded that permeability models, including Winland equation, produce erroneous results when their original coefficients are used. In contrast, the accuracy of predicted permeabilities improves with calibrated coefficients. It is worth mentioning that their database had nearly normal porosity and permeability distribution (Fig. 5.4). Higher degree of accuracy is achieved along with decreasing heterogeneity of the samples.

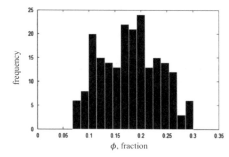
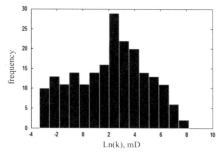

Fig. 5.4 Porosity and permeability distribution in 206 carbonate samples of Nooruddin database (Nooruddin et al. 2014)

5.2 Selecting the Parameters

Heterogeneity affects the petrophysical calculations, as discussed previously. So, the accuracy of formation evaluation results strongly depends on the appropriate selection of the parameters for these calculations. Two main approaches are used to obtain the reservoir characteristics of any hydrocarbon formation. They include deterministic and stochastic petrophysics. In deterministic approach, no randomness is involved in the calculations and results. A deterministic formula always yields the same result from a given starting condition or initial state. The user can change the formula or initial inputs. The result will be different, but it is always the same for the same assumptions and equations. Volume of shale (V_{sh}) calculation from gamma-ray (GR) log is a good example. In linear method, V_{sh} is calculated using the following formula (Eq. 5.4):

$$V_{sh} = \left(GR_{log} - GR_{min}\right)/(GR_{max} - GR_{min}) \tag{5.4}$$

where GR_{log} is the GR log reading at desired point, GR_{min} is the minimum GR value in the studied interval, and GR_{max} is its maximum. V_{sh} of a point is always the same using this equation. There are also some other formulas which can be used to calculate the V_{sh}, and so the result will be different but the value of this parameter is always constant using each formula. All of these formulas consider only a single formation property and use one or at most two (e.g., V_{sh} from neutron–density cross-plot) formation responses. A deterministic petrophysical approach uses a series of stepwise calculations. For example, V_{sh} is calculated based on the GR log data. Then, the shale effect is removed from the neutron responses using these data and porosity is determined. These two results are used to determine the volume of other minerals in the reading point. The effects of other parameters are ignored in these calculations. For example, the response of the neutron tool can be affected by gypsum mineral or the GR reading may be due to the K-feldspars or micas which are ignored in these calculations.

Some formation characteristics are selected in each step depend on the used formula. For example, for calculating the V_{sh}, minimum and maximum amount of GR readings are necessary. These constants may be simply the minimum and maximum of the GR data in whole interval, but the clay types may be different through the whole reservoir part. So, different zones with different clay types should be defined to achieve realistic values.

Wyllie equation for calculating porosity from sonic velocity data is another example. He related these two parameters with a linear equation which is known as Wyllie time–average equation (Eq. 5.5).

$$\Phi = (\Delta t - \Delta t_m)/(\Delta t_f - \Delta t_m) \tag{5.5}$$

where Φ is porosity, Δt is sonic transit time in true formation, Δt_m is sonic transit time in rock matrix, and Δt_f is sonic transit time in formation fluid.

Sonic velocity in carbonates depends on lithology, porosity, pore types and rock texture. It is faster in solid materials than in liquids. Dolomites have higher sonic velocity compared to calcite. Carbonate rocks with isolated and large pores (molds and vugs) also have higher velocities compared to the carbonates with the same porosity but containing small, connected and regularly distributed pore spaces. So, calculated porosities from the Wyllie's formula strongly depend on the rock properties including lithologies and pore types. Figure 5.5 shows the calculated porosity from Wyllie equation in Fahliyan Formation (Lower Cretaceous) in one of the Iranian southwest hydrocarbon fields. At first, porosities have been calculated using a constant vale for matrix sonic transit time (44 μs/ft) which is average value for limestone and dolomite (Fig. 5.5a). Then, porosities recalculated for each lithology with their related sonic transit time (50 μs/ft for anhydrite, 40 μs/ft for dolomite and 45 μs/ft for limestone) (Fig. 5.5b). It can be seen in this figure that results are clearly different. Calculated porosities of anhydrite samples are highlighted in green. These values considerably decreased when samples were classified according to their lithologies.

The problem is more complicated about the pore types. There is no known sonic transit time value for samples with different types of porosities. So, the resulted porosities should be corrected. Samples with different amounts of various pore types are tested in laboratory, and a correction factor is achieved.

Analyzing two logs in deterministic approach is also possible to gain a petrophysical parameter such as V_{sh}. Neutron–density (ND) cross-plot is used for this purpose (Fig. 5.6). In this method, the density of shale, matrix and formation water must be known. The neutron porosity (N_Φ) in a pure shale layer is also needed. The neutron response of the formation water is 1 (v/v) or 100 (%). The matrix, shale and water points are plotted in the ND cross-plot. These points are connected to each other, and a triangle is formed. V_{sh} is 1 (v/v) at the shale apex and zero on the water-matrix line. V_{sh} of any new point in then calculated with respect to its position. Bulk density (ρ_b) of the matrix can be gained by reading the ρ_b log at the zero porosity point. In other words, ρ_b is the same as matrix density (ρ_{ma}) when N_Φ is zero. In the same way, the shale density and porosity are gained by reading the ρ_b and N_Φ logs at the maximum GR reading, respectively. Different amounts of shale may be gained using various amounts of ρ_{ma}. So, for acquiring an exact amount of V_{sh}, the formation must be grouped into various zones with different lithologies. This is also the case for shale and formation water, but they are not as important as matrix because their density is not very variable in reservoir studies.

The situation is different for stochastic petrophysics. In this approach, the calculated value of each reservoir parameter (such as porosity, water saturation and lithology) depends on the other parameters, constants and formulas. All calculations are performed in a multidimensional space including various formulas for calculation of petrophysical parameters from different logs. So, the element of randomness is involved. In a stochastic model, several variables are unknown. So, if you change one variable in one equation, all results are changed. A group of response equations is constructed using all minerals and fluids volumes. Equation 5.6 shows the general form, and Eq. 5.7 shows an example for a suit of photoelectric factor (PEF) log

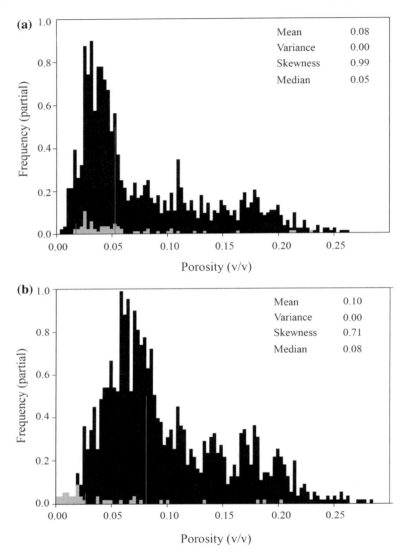

Fig. 5.5 Porosity distribution in Fahliyan Formation (Lower Cretaceous) in southwest Iran. See text for more explanations

responses in a carbonate formation. The rock is composed of two minerals including calcite and dolomite and two fluids, gas and water.

$$R_{\log} = P_1 V_1 + P_2 V_2 + \cdots + P_n V_n \tag{5.6}$$

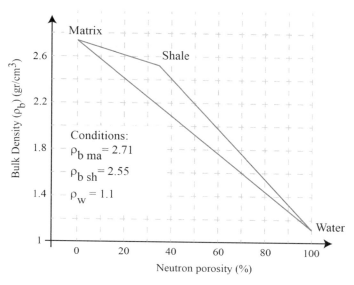

Fig. 5.6 Neutron–density cross-plot for calculating V_{sh} in carbonate formations. The position of the apexes may be different according to the shale, matrix and water densities and neutron responses

where R_{log} is predicted log response, P is log related property, and V is volume of that property.

$$PEF_{log} = PEF_{cal} V_{cal} + PEF_{dol} V_{dol} + PEF_{wat}(S_w F) + PEF_{gas}(S_g \Phi) \qquad (5.7)$$

where PEF_{log} is log reading at the desired point, V is volume, cal is calcite, dol is dolomite, S_w is water saturation, S_g is gas saturation, and Φ is porosity. It is worth mentioning that saturations have been multiplied by porosity because volumes of minerals are part of 1 while saturation is part of porosity. Then, the best match between the measured and predicted log responses (R_{log} in Eq. 5.6) is calculated using the least square method. The final error is calculated by summing the errors of predictions for all observations. The final model contains all volumes of minerals and fluids (Eq. 5.8).

$$\begin{bmatrix} p_{11} & p_{12} & p_{13} & p_{14} \\ p_{21} & p_{22} & p_{23} & p_{24} \\ & & & \\ & & & \\ & & p_{mn} \end{bmatrix} \cdot \begin{bmatrix} v_{11} \\ v_{21} \\ \\ v_{n1} \end{bmatrix} = \begin{bmatrix} r_1 \\ r_2 \\ r_3 \\ \\ r_{m1} \end{bmatrix} \qquad (5.8)$$

where p is standard log response for minerals and fluids (PEF_{cal}, for example), v is volume of the constituents, m is the number of used logs in the model, n is the number of constituents (minerals and fluids), and r is log response. A separate model is constructed for each point including different volumes and readings. Although the calculation method is completely different compared to deterministic approach, the variables are the same. Constants and matrix properties of each unit are different according to the reservoir heterogeneity. Several models should be constructed with different properties. Changing these properties in various models leads to different volumes of the constituents. While mineral properties (such as calcite density) are still constant, various zones of the reservoir have different minerals, pore types, pore volume, permeabilities and so on. For example, ρ_{ma}, GR and neutron responses of shale, matrix transit time, photoelectric factor and fluid types may change from one sample or zone to another. Several petrophysical programs have been developed which can apply different models to different zones or even different samples.

5.3 Solving the Problem

Selecting optimal constants and constituents has strong effect on the final results of petrophysical evaluations. This is exactly where reservoir heterogeneity should be taken into account. Different criteria may be used for managing heterogeneity of different parameters. While Archie's exponents strongly depend on pore types, the Wyllie's equation for calculating porosities in carbonates uses the matrix transit time which completely depends on lithology. So, rock typing or reservoir zonation should be based on similar pore types for calculating of S_w. In contrast, both lithology and pore types should be involved in rock typing for calculating porosity based on Wyllie's equation. So, the best rock typing method must integrate all reservoir properties within a single unit. This is one of the most challenging tasks in reservoir studies. Various rock typing methods should be applied to different data from diverse carbonate reservoirs of the world to clarify the best methods for each formula or model. Figure 5.7 shows porosity versus sonic velocity data for an Oligo–Miocene carbonate reservoir in south Iran. Four methods of rock typing have been applied to data including different lithologies (lime and dolomite), pore types (interparticle, moldic, intercrystalline and vuggy), ranges of permeabilities and various textures. The lines of Wyllie's equation for limestone and dolomite matrix are also illustrated. It is obvious that dolomite samples generally have higher sonic velocity (Fig. 5.7a). In low porosity values, data follow the Wyllie's equation. With increasing porosity, departures also increase. Samples with high porosity and permeability are far from the lines (Fig. 5.7b). Pore types of these samples include interparticle, moldic and intercrystalline in most cases (Fig. 5.7c). Both mud-dominated (mudstone) and grain-dominated (grainstone) samples are present in this part of the diagram (Fig. 5.7d). So, high porosity and permeability are responsible for such departure. These samples should be grouped in one rock type during the process of rock typing of this reservoir.

The final step is correlating the rock types with the wire-line log data. Logs are available from almost all wells and reservoir intervals and so can be used for correlation. Electrofacies (EFs) are useful for this purpose. Rock types are compared with EFs. Rocks with the same EFs in different wells have nearly the same properties.

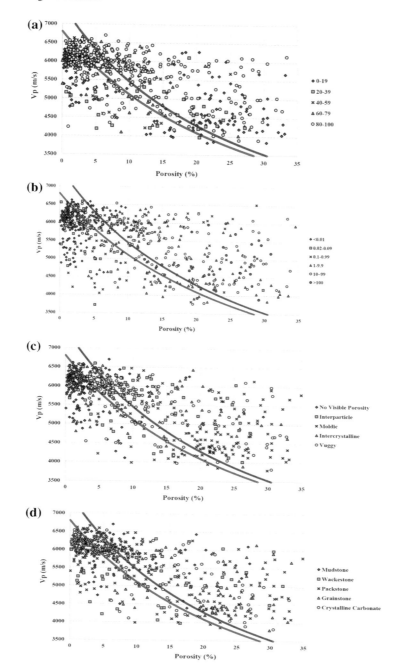

Fig. 5.7 Various rock typing methods and their relationships with Wyllie's equation for limestones (blue line) and dolomites (red line). Rock typing based on dolomite percent (**a**), range of permeabilities (**b**), pore types (**c**) and textures (**d**)

So, the same constants and coefficients can be used for their petrophysical evaluation. These EFs are used for the final static reservoir modeling which indicates the distribution of the properties in 3D space.

It should be mentioned that in all cases, the effect of overburden pressure must be involved in calculations. Both porosity and permeability decrease with increasing depth and pressure. So, the estimated oil in place is always more than the real value in the reservoir.

References

Kolodizie SJ (1980) Analysis of pore throat size and use of the Waxman-Smits equation to determine OOIP in Spindle Field, Colorado. SPE paper 9382 presented at the 1980 SPE Annual Technical Conference and Exhibition, Dallas, Texas

Lucia FJ (2007) Carbonate reservoir characterization: an integrated approach. Springer, Berlin, Heidelberg

McPhee C, Reed J, Zubizarreta I (2015) Core analysis: a best practice guide. Elsevier, London

Nazemi M, Tavakoli V, Rahimpour-Bonab H, Hosseini M, Sharifi-Yazdi M (2018) The effect of carbonate reservoir heterogeneity on Archie's exponents (a and m), an example from Kangan and Dalan gas formations in the central Persian. Gulf J Nat Gas Sci Eng 59:297–308

Nooruddin HA, Hossain ME, Al-Yousef H, Okasha T (2014) Comparison of permeability models using mercury injection capillary pressure data on carbonate rock samples. J Petrol Sci Eng 12:9–22

Nooruddin HA, Hossain ME, Al-Yousef H, Okasha T (2016) Improvement of permeability models using large mercury injection capillary pressure dataset for middle east carbonate reservoirs. J Porous Media 19(5):405–422

Tavakoli (2018) Geological core analysis: application to reservoir characterization. Springer, Cham

Tiab D, Donaldson EC (2015) Petrophysics, theory and practice of measuring reservoir rock and fluid transport properties. Gulf Professional Publishing, Houston

Printed in the United States
By Bookmasters